Healthcare Emergency Incident Management Operations Guide

Healthcare Emergency Incident Management Operations Guide

Jan Glarum
Independent Consultant, Chicago, IL, United States

Butterworth-Heinemann
An imprint of Elsevier

Butterworth-Heinemann is an imprint of Elsevier
The Boulevard, Langford Lane, Kidlington, Oxford OX5 1GB, United Kingdom
50 Hampshire Street, 5th Floor, Cambridge, MA 02139, United States

Copyright © 2017 Elsevier Inc. All rights reserved.

No part of this publication may be reproduced or transmitted in any form or by any means, electronic or mechanical, including photocopying, recording, or any information storage and retrieval system, without permission in writing from the publisher. Details on how to seek permission, further information about the Publisher's permissions policies and our arrangements with organizations such as the Copyright Clearance Center and the Copyright Licensing Agency, can be found at our website: www.elsevier.com/permissions.

This book and the individual contributions contained in it are protected under copyright by the Publisher (other than as may be noted herein).

Notices
Knowledge and best practice in this field are constantly changing. As new research and experience broaden our understanding, changes in research methods, professional practices, or medical treatment may become necessary.

Practitioners and researchers must always rely on their own experience and knowledge in evaluating and using any information, methods, compounds, or experiments described herein. In using such information or methods they should be mindful of their own safety and the safety of others, including parties for whom they have a professional responsibility.

To the fullest extent of the law, neither the Publisher nor the authors, contributors, or editors, assume any liability for any injury and/or damage to persons or property as a matter of products liability, negligence or otherwise, or from any use or operation of any methods, products, instructions, or ideas contained in the material herein.

British Library Cataloguing-in-Publication Data
A catalogue record for this book is available from the British Library

Library of Congress Cataloging-in-Publication Data
A catalog record for this book is available from the Library of Congress

ISBN: 978-0-12-813199-2

For Information on all Butterworth-Heinemann publications
visit our website at https://www.elsevier.com/books-and-journals

Working together
to grow libraries in
developing countries

www.elsevier.com • www.bookaid.org

Publisher: Candice Janco
Acquisition Editor: Amy Shapiro
Editorial Project Manager: Emily Thomson
Production Project Manager: Vijayaraj Purushothaman

Typeset by MPS Limited, Chennai, India

DEDICATION

This book is dedicated to all my current healthcare clients. Their dedication to enhancing the preparedness of their organization reminds me that we can do better when the unthinkable happens.

After 40 years of experience in emergency medical services, law enforcement, fire service, emergency management, and healthcare I think the topic covered in the book is critical. I never worry if operations-level personnel will do their job. What I have seen and what I worry about is poor incident management. The result of which is not enabling operations to do their job either safely or efficiently. At best poor incident management wastes resources. At worst, it hurts or kills people needlessly.

There is a reason this book is so short. The Incident Command System (ICS) is not and should not be difficult. I wanted this book to allow someone to put on a class and use the need to know material. I also wanted this book to serve as a reference for initial incident management. I hope by the time readers breeze through the book they will have a better skill set to revise, customize, and practice using the ICS within their organization.

Regardless of any new tool you pick up, there is a learning curve. Play with the ICS, use it to prepare for the next preplanned event in your community. Practice, practice, and practice as incident management skill sets are perishable. I cannot tell you when it will happen or to whom, but someday one of my readers will need to implement the ICS to manage a major event. I truly believe if you make an investment in learning and using the ICS it will change the outcome for someone.

CONTENTS

Introduction ix

**Chapter 1 Overview of National Incident Management
System and Hospital Incident Command System** 1
National Incident Management System 1
Incident Command System 2
Hospital Incident Command System 6
Incident Command System Characteristics 7
Incident Command System Organization 11

Chapter 2 Command and Unified Command 17
Single Command 17
Unified Command 18
Command Staff 21
Agency Executive 23
Command Center 24

Chapter 3 The Operations Section 27

Chapter 4 The Planning Section 33

Chapter 5 The Logistics Section 35

Chapter 6 The Finance/Administration Section 39

Chapter 7 Intelligence/Investigations 41

Chapter 8 Area Command 45

Chapter 9 Incident Action Planning Process 51
Relationship Between Your Emergency Operations
Plan and an Incident Action Plan 51
Understand the Situation 52

Determining Your Operational Period 54
Establish Incident Objectives and Strategy 55
Develop the Plan 59
Prepare and Disseminate the Plan 59
Execute, Evaluate, and Revise the Plan 60
The Planning P 61
Forms 63
Keys to Success 66

Chapter 10 Hospital Incident Command System Initial Incident Action Plan Library for Common Events 67
Initial Incident Action Plan Form—Active Shooter Event 69
Initial Incident Action Plan Form—Earthquake Event 71
Initial Incident Action Plan Form—Flooding Event 73
Initial Incident Action Plan Form—Ice Storm 75
Initial Incident Action Plan Form—Loss of Internet Service
(Including VoIP Phones) 77
Initial Incident Action Plan Form—Tornado Watch/Severe
Wind Event 79
Initial Incident Action Plan Form—Tornado Warning Event 81
Initial Incident Action Plan Form—Tornado Event 83
Initial Incident Action Plan Form—Loss of Water Event 85
Mini Exercise List 86

Additional Reading 91

INTRODUCTION

I wrote this book for one simple reason—I want readers to use the Incident Command System (ICS). I did not use the term Hospital Incident Command System (HICS) on purpose. Why? Because unlike the HICS, the ICS is designed to be discipline blind. I think that makes it easier to learn, apply, and remember when it is needed most.

Whether you are managing the response and recovery of an emergency, conducting a class, or facilitating any preplanned event such as a mass vaccination campaign, the ICS should be your first choice. That is because, at its core, the ICS is a proven and effective management tool designed to manage all manner of resources, in a wide range of situations. If you need it to, the ICS could even be used to manage a wedding ceremony.

It is possible to master the basics of the ICS quickly. Why waste time trying to reconfigure your way of doing business? The ICS allows flexibility to design position titles that make sense for your organization and eliminate filling out forms that are foreign to your day to day operations.

> Let me share an example. For many years, I organized free, Emergency Medical Services (EMS) continuing education for volunteer fire and ambulance services located in rural Oregon. I started the program when I administered a hospital-based ambulance service, and continued even after I left to pursue work as an Emergency Manager.
>
> Oregon has areas designated as frontier, which means prehospital response can take up to 2 h. There are not many people living there, virtually no local paramedics, and the closest hospital is over 2 h away. Ambulances run less than one-hundred calls a year—and when those calls come, they can be serious. Because the volunteers did not see many calls, I tried to work in as many scenarios as possible during our weekend sessions.
>
> One year, they asked for ICS training in addition to EMS topics. I assembled a small team which included myself, an accident reconstruction expert from the Oregon State Police, and the fire chief from the volunteer fire department who was also an EMT-Intermediate. We were given a portable radio when we arrived just in case an ambulance call happened.

Unfortunately, early on the last morning of training, a call came in of a reported one vehicle rollover very close to the border with Idaho. I woke my team members and got everyone piled into their respective cars. The fire department was also dispatched since they carried all the extrication tools.

I asked the fire chief, "Do you think they set this scenario up just to get back at all the ones we threw at them?" Unbeknown to us, the crew in the ambulance was thinking the same thing, "I bet Jan set this up!"

As we neared the scene I contacted the fire department by radio and asked them to establish the ICS when they got on scene. They had their first ICS class the day before, and this would be a good test for them.

The fire department got on scene before we did and announced they were establishing command. Command advised there was one vehicle involved with two ejected patients. Command asked me to secure the west end of the scene with my vehicle and directed the state trooper to secure the east side with his and handle the law enforcement needs. They directed the ambulances to park near the patients and asked someone from the ambulance to act as the Medical Branch Director.

The Medical Branch Director assigned one ambulance crew to each patient—both were conscious and significantly injured. I joined the crew with the more critical patient and the EMT-Intermediate fire chief joined the other. We assessed the patients and reported back to the Medical Branch Director that both patients needed to get to a trauma center. The Medical Branch Director checked on air ambulance availability since the closest Level 3 trauma hospital was almost 3 h away by ground. About the time our patients were packaged up the Medical Branch Director advised us no air ambulance was available due to weather. I asked the Medical Branch Director to let the hospital know that we were on our way with two trauma entries.

As we loaded the patients into the ambulances, the Medical Branch Director poked her head in the back of my ambulance. She was beaming and told me the hospital was notified. I asked why she was so excited and she said that the hospital never activates the trauma team for them, but this time they did. I told her it was because her report was organized, to the point and left no doubt what we were expecting the hospital to do for these patients.

As we were pulling away from the scene of 15 responders, 2 fire trucks, 2 ambulances, a sheriff's department vehicle from a county five-hundred miles away, and a state police accident reconstructionist; a local deputy pulled up. Command was transferred over to the local deputy as the role of EMS and the fire department were completed. The deputy took over the investigation with the assistance of the state police trooper. My trooper friend told me later the deputy said they had never seen such a well-organized scene in the county before. Nor did he ever have so much help arrive.

Both patients were flown out of the Level 3 to a Level 1 in Idaho several hours later. The entire scene that involved multiple agencies and jurisdictions was handled smoothly by a team that had received practical ICS training just the day before. Two people filled key roles in managing this scene—the Incident Commander (IC) who was responsible for the entire incident and the Medical Branch Director who supervised all medical activities and coordinated back through the IC.

The ICS does not require much except a good working knowledge of the tool. ICS worked in this emergency and in many others I have been involved with over the years. It will work for you, too.

I think you will find this book accomplishes a couple of things:

It is a fun read. This may sound counterintuitive for a book about the ICS, though I ask you to humor me. There are plenty of dry ICS textbooks. You will find this one is something different.

This book can serve as a course guide for conducting in-house education and exercises. In the end, you will be able to apply critical thinking skills and be competent using the tool. If you want to "pass a test," this is not really the book for you. If you want to be organized and effective the next time you have to manage an emergency, read on.

You will find the library I have built of Initial Incident Action Plans to be useful in real events. They may not fit the situation perfectly, though I find it is better than starting from scratch. I hope this will be a dog-eared, well-thumbed book after a couple of years.

This book eliminates the phobia around using the ICS. By simplifying, you can make the tool work for your organization's way of doing business rather than trying to conform your system to a rigid series of position titles and forms.

If you are familiar with ICS, you can skip around and find the chapters that will help you the most. To help you determine where to start dog-earing pages, here is a summary of each:

Chapter 1, Overview of National Incident Management System and Hospital Incident Command System, provides the history of why we are voluntarily mandated to adopt National Incident Management

System (NIMS). The background information might be useful in getting the C-Suite involved in ICS adoption. This chapter also includes a quick review of the key principles and features of the ICS and the HICS. Check out this chapter if you want some tips on blending naming conventions within your organizational chart.

Chapter 2, Command and Unified Command, covers my views on command options and staff positions, including the balance between IC and agency executives. I will discuss the role a Command Center plays and why going virtual may be a great option.

Chapter 3, The Operations Section, offers various options on structuring the personnel assigned to accomplish work. ICS terms have specific meanings and become critical when you need to work with other organizations on larger events.

Chapter 4, The Planning Section, covers what I think is one of the most underutilized functions within the ICS—planning. The right planning person with the best plan is key to preventing unpleasant surprises during an event.

Chapter 5, The Logistics Section, looks at how to get all the "stuff" needed to support the organization. The Communications Unit normally lives here, though I think a smart IC needs to elevate the function in stature. I will state my case.

Chapter 6, The Finance Section, is often an afterthought. It should not be. If the event is big enough, you may be eligible for some type of compensation. If you are on a tight budget or hope to turn a profit, you will want to review this chapter.

Chapter 7, Intelligence/Investigations, is a relatively new comer to the ICS organizational charts. This is more than a law enforcement function and has several options for utilization in your organizational structure.

Chapter 8, Area Command, may not need to be utilized very often in your region, though if it is, you better understand the ins and outs of this role.

Chapter 9, Incident Action Planning Process, gets into material that will determine the course of response and recovery, good or bad. Normally covered in the ICS-300 course, many instructors bury students in forms instead of valuing the process. Me? I hope to jump-start your new love of incident action planning.

Chapter 10, Hospital Incident Command System Initial Incident Action Planning Library for Common Events, offers templates for a wide variety of threats and hazards to your organization. The list was crowd sourced through hospital emergency management personnel on LinkedIn and reflects a broad range of international and domestic opinions. I hope this will become one of the most used chapters in the book.

Spoiler alert: I do not include an IAP for meteor strikes. I know this will come back to haunt me one day—a meteor passed over Chicago as I slept on February 6, 2017.

Chapter 10, Mini-Exercise List, provides ideas on exercises. These cover a variety of natural, technological, and human-caused events. I have each exercise icon-marked for quick reference according to the relative time, money, and staffing commitments required.

Ultimately, I want you to understand why you should adopt the ICS and get good at using it. Why? I have participated in too many hospital exercises where unnecessary effort is expended trying to learn the HICS fresh out of the box. Over time I expect more to follow and this book will help your organization become one of them.

CHAPTER 1

Overview of National Incident Management System and Hospital Incident Command System

NATIONAL INCIDENT MANAGEMENT SYSTEM

On September 11, 2001, 19 militants associated with the Islamic extremist group al-Qaeda hijacked four airliners and carried out suicide attacks against targets in the United States. Within 2 weeks, Pennsylvania Governor Tom Ridge was appointed as the first Director of the Office of Homeland Security and brought the Department of Homeland Security (DHS) to life within a year. Today, there are over 240,000 DHS employees. The DHS trails only the Department of Defense and Department of Veteran Affairs in sheer size.

Director Ridge was tasked to develop, among other things, a National Incident Management System (NIMS). The goal was to provide a consistent nationwide approach for Federal, State, and local governments to prepare, respond, and recover from incidents impacting the homeland.

In 2005, to add some teeth into NIMS, Federal agencies made adoption of the NIMS a requirement for providing Federal preparedness assistance through grants or contracts. Organizations that accept Federal grants across the country officially adopted NIMS in some type of proclamation or declaration. No one wanted to lose the access to grant dollars. If only this new level of coordination was so simple.

The first problem with NIMS resolution adoption activity? Agencies and organizations who claim to be NIMS compliant often fail to institutionalize the components of the NIMS much less use these components when they are needed. They set themselves up for a fall. NIMS implies that these organizations will use the Incident Command System (ICS) to manage a major event. If there is a bad outcome and it is determined that failure to use the ICS was a contributing factor, it

Figure 1.1 Training scenario where students spent 5 days using the ICS in managing difficult events.

may expose organizations to litigation, fines, and review of any federal funding that is been received (Fig. 1.1).

The second problem with NIMS is that the initial management of any event is handled at the local level where NIMS adoption is lower. When an event gets big enough to exceed local or even state capabilities and more assistance is needed, then the federal government comes into help using NIMS. Sometimes the arrival of the cavalry results in a local incident management system that faces a whole new set of challenges.

INCIDENT COMMAND SYSTEM

The ICS is a key feature of the NIMS.

With every major event, we seem to have the same problems when it comes to managing the consequences. The 1993 World Trade Center attack, 1995 bombing at the Alfred P. Murrah Federal Building in

Oklahoma City, 2001 World Trade Center attack, Hurricane Katrina in 2005, and Hurricane Sandy in 2012 each faced issues with:

- command and control
- communications
- resource management

It should come as no surprise that the ICS was selected to be a key component of NIMS. The ICS is all about command and control, communications, and resource management.

NIMS does not tell an Incident Commander (IC) what that the organizational structure should look like, what the IC should try to achieve, or how to go about it. But with preplanning, it is possible to develop a guide that will help an IC to get an event organized in the early, critical moments. If you are interested in these plans, flip ahead to Chapter 10, Hospital Incident Command System Initial Incident Action Plan Library for Common Events, to review a library of guides. These can be customized for use by your organization. The ICS is a management tool with some absolutes but a great deal of flexibility for events that are simple to complex, planned and unplanned, small or large (Fig. 1.2).

By using the ICS for scheduled events, you will gain the experience and skills needed to confront an emergency. Soon, you will quickly identify the best way to go about getting the first operational period organized and underway.

Figure 1.2 Hospital decontamination operations are perfect example for using the ICS.

We are supposed to use the ICS in order to be NIMS compliant, but this does not mean everyone must operate under one massive umbrella. ICS is based on the premise that if the IC opens a new function, it needs to be staffed by someone trained and competent in carrying out the duties of that function. The larger you build your structure, the greater the need for trained incident management personnel.

So, who needs to be trained and how many people should be involved? A small organization might imagine that it can get by with minimal training. That is not quite the case. The following chart lists the baseline ICS courses up to the ICS 200. Advanced training is a term given to the ICS 800, ICS 300, and ICS 400 (Fig. 1.3).

The need for the advanced training courses is based on the complexity of the incident an organization can reasonably expect to see. The next chart explains the different incident complexity levels. Even though you may have a small organization, if you are at risk for a Type 1, 2, or 3 event, you will need to obtain the more advanced ICS education (Fig. 1.4).

The incident complexity level is a great way for an organization to capture in policy when to expect the ICS to be used and to what extent, e.g., require a written Incident Action Plan (IAP) for any Type 3 event or higher in complexity. You can then go on to describe

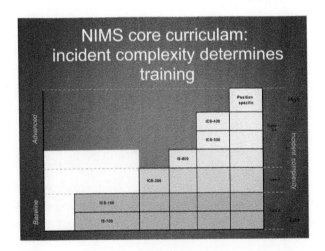

Figure 1.3 Chart showing relationship between ICS training and NIMS event typing.

Overview of National Incident Management System and Hospital Incident Command System

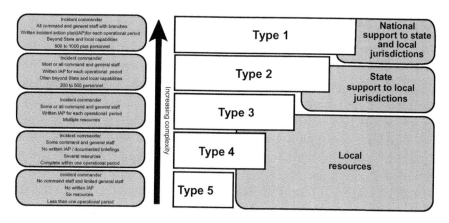

Figure 1.4 Chart showing relationship between incident complexity, ICS features used, and levels of governmental support.

examples of each complexity level in your own words to make it easier to train personnel when to recognize each level. I always help my clients clearly articulate these levels and put it in policy that the ICS will be used. It helps ensure an organization will be NIMS compliant in deed and not just in words.

In addition to allowing a customized organizational structure to be built the ICS also guides the process of planning. One of the issues with the current ICS course objectives is that students do not receive any depth of information on the ICS planning process until the ICS 300 level. I introduced the planning process in my advanced Hospital Incident Command System (HICS) courses years ago to address this deficiency. In Chapter 9, Incident Action Planning Process, I will offer the reader everything you need to know to understand this key feature of the ICS.

The ICS can be used by all levels of government, public and private organizations, and any volunteer group that wants to be more efficient in their operations. The ICS has six functional areas: command, operations, planning, logistics, finance/administration, and intelligence/investigations. The Intelligence/Investigation section is a relatively new addition to the traditional ICS structure. This section has a good deal of flexibility in where it can be placed in the organizational chart—I will address that in more detail in Chapter 7, Intelligence/Investigations. There are no rules as to how many functions need to be open. The organizational structure should be tailored to the event.

HOSPITAL INCIDENT COMMAND SYSTEM

The ICS was developed in the 1970s following a series of catastrophic fires in California's urban interface. Property damage ran into the millions, and many people died or were injured. The ICS met the needs of the fire-fighting community, which was reasonable. However, spreading the use of ICS outside of that application caused confusion. Initial ICS courses were taught by fire personnel since they had the most experience using the tool. Unfortunately the material was not always presented as discipline blind and was heavily skewed toward a fire or hazardous materials world. This created some doubt as to whether to use the ICS among nonfire organizations.

The California Emergency Medical Services Authority first released the Hospital Emergency Incident Command System (HEICS), in the late 1980s. The intent was to present the ICS with a hospital focus. They have revised this system a number of times—the latest in 2014. Job action sheets for every imaginable position that might be needed in the hospital environment were created. Part of the idea behind developing these job action sheets was to allow a just-in-time utilization by hospital staff as most do not use the ICS daily like fire service personnel do. While the job task sheets, along with the incident planning guides and incident response guides, that were developed can be useful tools, most will need to customize them for an organization.

The difficulty with the job action sheet concept is that the ICS is built on the premise that personnel assigned to any specific functional area have been trained in how to perform the duties associated with that function. Imagine a trauma team activation is called at your hospital. Instead of each person arriving to perform specific duties based upon their skill set, you hand out job action sheets based on the order they enter the Emergency Department. The first person to arrive will be the anesthesiologist, the second the trauma surgeon, followed by imaging, and then the trauma nurse. Instead of assigning everyone to known and understood responsibilities the attention is placed on unfamiliar job action sheets. Throw in foreign-looking HICS forms and make staff wear a colored vest so everyone by function, and you will have a recipe for chaos. I know a number of my readers are laughing—they have been there, done that. The point is that positions on your Incident Management Team are just as important as those on your trauma team. Each brings a specific skill of perishable skills if not maintained.

What I find when I work with facilities who have bought into the packaged HICS system is that personnel spend more time trying to figure out how to do their HICS assigned job and fill out forms than to manage the response to the event. There are a couple of solutions. First, take the time to customize the generic HICS to conform to your hospital's way of doing business, not the other way around.

Job action sheets are a great place to start, but each organization needs to determine for themselves what functions are most likely to be filled and the duties they will perform. Or, you can do as I suggest—drop the H from HICS and simplify. The ICS is discipline blind by design. I have seen public health, public works, law enforcement, hospitals, and other organizations put their discipline initials in front of ICS and think they have solved something. All this does is make using the ICS harder than it needs to be.

INCIDENT COMMAND SYSTEM CHARACTERISTICS

There are 14 established characteristics which contribute to the success of the ICS as an incident management tool. These can be used effectively by anyone interested in making the effort to learn and practice.

1. Common terminology. The ICS uses common terminology and avoids the use of codes. Hospitals are famous for using codes even though there are no agreed upon terms that are used in the industry. On day to day matters different terminology is not an issue. However, during NIMS Type 3, 2, or 1 events where organizations assist others and/or receive assistance, it can be a huge issue. The value in moving to common terminology in day to day use is that it allows staff to seamlessly work with others during emergencies. To be fair the problem of not using common terminology is not restricted to healthcare organizations.

> In the fire service, it is common to label the sides of a building A, B, C, and D. Law enforcement often uses 1, 2, 3, and 4 to denote the same sides. Some federal partners may use colors to identify sides. It is not hard to see how this could be an unnecessary disaster if an event with fire, law enforcement, and federal partners. Despite best efforts, getting

everyone to speak the same language is never going to happen. However, there is a solution for use during major events. I will discuss it in Chapter 2, Command and Unified Command, where Unified Command can be found.

2. Modular organization. The ICS organization is built by opening the Incident Command function or box. The ICS organization can simplistically be called a series of sticks and boxes. Boxes indicate a specific functional area to be opened and staffed, and sticks represent the lines of supervision. The Incident Command box may be the only one that needs to be used during small events. As the incident grows in complexity, scope, or hazards, additional boxes may be opened. This allows the IC to divide event management into logical functional areas and delegate the performance of those functions to qualified staff.
3. Management by objectives. The IC is responsible for establishing the objectives to be accomplished during an operational period. Once the objectives are established, strategies can be developed to meet the objectives. Strategies lead to specific tactics and tactics lead to tasks that need to be given to the people trained to carry them out. Proper preplanning to identify the tasks that will need to be carried out during specific event management can help identify training needs. By knowing what needs to be accomplished over a set period of time and by whom the IC can measure progress and make adjustments as necessary. This feature helps ensure efficiency in resource utilization and effectiveness in operations.
4. Incident action planning. An IAP provides a concise, easy to understand means of communicating the overall incident management priorities, objectives, strategies, and tactics in the context of operational and/or support activities. I discuss this IAP process in detail in Chapter 9, Incident Action Planning Process.
5. Manageable span of control. Anyone who has taken an online ICS course knows the correct test answer for span of control questions is three to seven people for every supervisor, with five being optimal. There is no magic number. The complexity or hazards associated with an operation is a better rule to follow. School teachers seem to do fine with a 25:1 ratio while a dive team might operate with a 2:1 ratio. The objective of a manageable span of control is to ensure one person can reasonably manage and

communicate with a specific number of subordinates. If too many people are reporting to one person, there will be unavoidable delays in communications coming in and going out. The IC can open another box and alleviate span of control issues by adding another supervisor.

6. Incident facilities and locations. If necessary, various support facilities may be established to help support event management. These could include a Command Center, Agency Operations Center, Decontamination Corridor, Alternate Care Site, mass casualty triage location, or vaccination/pharmaceutical points of distribution, and others as required.
7. Comprehensive resource management. To maximize efficiency and effectiveness of operations, it is critical to have a real-time handle on resources assigned and available for utilization. Resources could be personnel, equipment, supplies, and facilities. Not currently seen in a hospital environment is a Passport Accountability System commonly used by fire services. It allows real-time accounting of all personnel operating within an incident. It was recently adopted by the Oregon Health Authority for use with the Medical Reserve Corps and the Oregon Disaster Medical Team and in exercises has proven valuable.
8. Integrated communications. Interoperable communications have been a buzzword for some time now in federal grants. Communication continues to come up as an issue in events and exercises of all sizes. It has been my experience that more radios do not result in better communication. Conversely, if command cannot communicate with someone, they certainly are not able to manage them. With the evolution of digital communication devices, organizations need to assess how best to assemble a communications network that meets the operational needs during a major event.
9. Establishment and transfer of command. Readers probably assume that the Command box will automatically be established. This is not always the case, particularly with personnel who do not routinely utilize the ICS. It needs to be documented in policy that the first action is to clearly establish command and ensure it is widely communicated. With my hospital-based ambulance service, I established a policy that anytime we had greater than three ambulances or two agencies responding to the same incident, we would establish Single Command or Unified Command as appropriate.

Command may be transferred at any point, but typically occurs when a new operational period begins. When command is transferred, there must be a briefing from the outgoing to the new IC to ensure all essential information concerning current activities and resources committed are communicated.
10. Chain of command and unity of command. Chain of command refers to the orderly line of authority within the ICS positions. Unity of command means that all individuals have one designated supervisor whom they report to within the ICS. This is required to eliminate the confusion that results from multiple, or conflicting orders. People tend to naturally look to their day to day supervisor for questions and assignments. However, when the ICS is activated, this may not be the case. Individuals in ICS supervisory positions will not necessarily be in a day to day supervisory position. This requires discipline within the organization to adhere to the ICS chain of command. Personnel working within the ICS organizational structure *and* the day to day supervisors must direct questions back to the appropriate ICS supervisor.
11. Unified Command. Anytime there is more than one jurisdiction involved in an incident and/or multiple agencies, it is good to establish Unified Command. An organization needs triggers for when Unified Command should be established in the same way they need them for the establishment of the ICS.
12. Dispatch/deployment. The just-in-time IAP development feature of the ICS allows for a tremendously efficient utilization of resources. The IC should only require those resources that are necessary to accomplish the objectives for the operational period. There is no reason to have hospital personnel report to the cafeteria to stand by. Hospitals are too lean in staff. If staff are needed, they will be requested. If not, they should continue with their normal duties.
13. Accountability. Effective accountability is critical to ensure safety and operational efficiency. The following principles must be observed:
    Check-in/check-out. All personnel assigned to the operation must report in and receive their assignment according to the process established by the IC.
    IAP. Operations must be carried out and in-line with accomplishing the objectives articulated in the IAP.

Unity of command. Each person assigned to incident operations must report to only one supervisor.

Span of control. Supervisors must be able to adequately communicate with and manage subordinates working for them.

Resource tracking. Supervisors must record and report any status in resources deployed or required up the chain to their supervisor.

14. Information and intelligence management. The IC must establish a process for gathering, analyzing, assessing, sharing, and managing incident-related information and intelligence.

INCIDENT COMMAND SYSTEM ORGANIZATION

The ICS is made up of five major functional areas. Command, operations, planning, logistics, and finance/administration. A sixth functional area, intelligence/investigations may be added if the IC choses. You will notice in Fig. 1.5 the five major areas are visible as well as a sixth titled social media monitoring and response. You can add and subtract functions as dictated by the event and your operational objectives.

Modular expansion. The ICS organizational structure is modular and designed to be built out starting from the Incident Command function. The IC adds the functional areas and elements necessary to accomplish objectives for a set period of time. The structure can remain very small and build as the scope, size, type, and complexity of the event dictates. Conversely, as objectives are accomplished, the organizational chart can be collapsed as functional areas are no longer

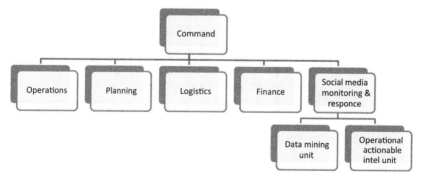

Figure 1.5 Organizational chart showing customization of functions based upon incident needs.

needed. Hospitals often can be overwhelmed when looking at commercially available HICS organizational charts. See the one shown in Fig. 1.6. This often leads to the opening of too many functional areas, at least initially. They end up with a top heavy organizational structure staffed with inexperienced supervisors and too few personnel to actually accomplish work.

Based upon event dynamics and need to maintain a manageable span of control the IC can open one or all the separate sections (operations, planning, logistics, finance/administration, and information and intelligence management). Yet another option is to develop a customized section and staff it appropriately to meet an identified need in event management.

If the IC opens a section, it will be led by a Section Chief who may further delegate supervisory authority for their functional area as required. A Section Chief may establish branches, groups, divisions, or

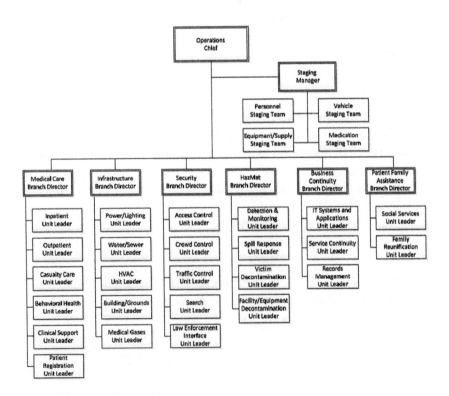

Figure 1.6 HICS organizational chart.

units. In turn, Unit Leaders may assign individual tasks within the unit.

Deputies and assistants. The use of deputies and assistants can be very useful to an IC if the situation requires them. Deputies may be from another organization if it makes operational sense based on event operations, responsibilities, or resource commitments. Deputies may also be used at the section and branch levels of the organization. Any deputy assigned to the incident should be qualified to fulfill the position as the primary supervisor.

Assistants are used to support Command Staff members. Command Staff consist of the Public Information Officer, Safety Officer, Liaison Officer, and Technical Specialist.

Building out the modular organization using the positions described above should be based on the following considerations:

Keep the organization as small as possible. Determine the incident objectives before building the organizational chart. If you open a functional area, it should directly relate to accomplishing or supporting the objectives for that operational period.

Opening a functional area comes with the expectation that the person staffing it has the appropriate level of training and experience to manage the area of responsibility.

If a functional area is not opened and staffed, the responsibilities for performing those functions rests with the next highest supervisory level position. For example, if the IC chooses not to open any other functional area, they are responsible for carrying out the functions of all those unopened positions.

Ensuring span of control is maintained at a manageable level. If the number of people reporting to one supervisor approaches more than six, consider the need to open a new functional area and divide up the work load.

Position titles are part of the ICS. Keeping titles consistent with traditional ICS terminology is a good idea in most cases. Don't be afraid however to change a position title to better reflect the function performed in your organizational setting. Keep in mind that NIMS/ICS says that you need to use an incident management system with features similar to the ICS. For example, if you open your hospital command center, you may call the person in charge the IC. You could also call them the command center director, supervisor, or executive if that makes more sense to your existing organizational structure.

Blending positions. This concept was included with the 2014 HICS revisions and is a great step forward, if properly conducted. In Chapter 10, Hospital Incident Command System Initial Incident Action Plan Library for Common Events, I give the reader an exercise to run which brings the lesson home. In many organizations, staff routinely perform multiple job roles daily. During an emergency, people will need to assume more than one position on the incident management team, at least initially.

The concept of blending is perfect for critical access hospitals, smaller organizations, or the third shift in most facilities. Some guidance was given with the 2014 HICS revisions on the idea of blending roles into single job action sheets but it is not as easy as stapling two or three job action sheets together and calling it good. Once you complete the exercise in Chapter 10, Hospital Incident Command System Initial Incident Action Plan Library for Common Events, you will have a better understanding of the logical way to blend positions so it fits naturally with positions staff have now.

At this point the reader has been reminded how we got to this NIMS/ICS and HICS place. You do need to be NIMS compliant and it must be more than lip service. The Department of Health and Human Services requires and expects hospitals and healthcare systems receiving Federal preparedness and response grants, contracts, or cooperative agreements to meet the following activities:

Organizational adoption of NIMS
Command and management
   ICS
   Multiagency Coordination System
   Public Information System
Preparedness planning
   NIMS implementation tracking
   Preparedness funding
   Revise and update plans
   Mutual-aid agreements
Preparedness training
   IS 700 NIMS

IS 800a NRP
ICS 100 or equivalent and 200 or equivalent
Preparedness exercises
   Training and exercises
   All hazard exercise program
   Corrective actions
Resource management
   Response inventory
   Resource acquisition
Communication and information management
Standard and consistent terminology

It may look daunting, but it is not. Adopt NIMS and the ICS comes along with it. Revise your existing emergency operations plans and standard operating procedures to reference that your organization uses the ICS to manage events. This allows you to keep your plans at the strategic level since ICS includes a real-time planning component. You can improve your chances of success by using the Initial IAPs found in Chapter 10, Hospital Incident Command System Initial Incident Action Plan Library for Common Events, or develop your own.

Enable your staff to receive ICS training that gives them the skill sets necessary to be competent. The standards listed above say ICS 100 and ICS 200 or equivalent. I conduct a lot of ICS courses for healthcare and other organizations that focus on doing just that. Identify staff for positions that closely match their existing skill set and personality. If you find yourself a good Planning Section Chief, lock them into a long-term contract. They will be worth their weight in gold to your organization.

Over the next few years there will be more catastrophic events will occur. All the money spent on NIMS will not determine the outcomes, good or bad, in the early stages of managing the consequences. That rests with your organization and the relationships you have built. ICS allows you all to work together, though it only works if we all agree to use it.

CHAPTER 2

Command and Unified Command

SINGLE COMMAND

When a trauma patient is brought into the Emergency Department, a team approach is used to manage the patient. One person is normally in charge of this effort. When a mass casualty event occurs and multiple patients are arriving at the hospital, supervisors are assigned to specific functions and staff deploy under their supervision in teams to perform tasks. These are examples of using the Incident Command System (ICS) on a regular basis in a hospital setting but not calling it by name.

So, what is the value of using the ICS rather than just go about business as usual? The value is in the repetition of use which builds confidence in the process. Trauma teams work and Mass Casualty Incidents (MCI) plans work because staff function in them enough they become second nature. Watching most hospitals trying to utilize Hospital Incident Command System in managing an event and it is anything but second nature. The challenge occurs when an unusual or low-frequency event occurs and staff do not have the muscle memory to carry them through.

The other key issue is that normal operations must go on within the organization. Using existing supervisory staff to continue normal duties and also manage a major event can be very disruptive to both daily operations and major event ICS functions. By utilizing the ICS the Incident Commander (IC) is given authority to manage the event, not the rest of the hospital, or concurrent with their normal duties as well. The IC does not get to write checks or supervise anyone not working within their ICS organization. The IC is responsible for determining objectives, building, and staffing the organizational structure to accomplish those objectives and resolving the event. Having one individual designated as the IC establishes the chain of command and unity of command to let this occur effectively and efficiently.

UNIFIED COMMAND

I was one of the trainers selected to go to Oklahoma City and provide response to terrorism events as part of the Soldier Biological Chemical Command program. I was very excited about the opportunity to go as this was not too long after Timothy McVeigh had detonated his device. I wanted to learn all I could from the responders to the bombing event. My excitement started to fade as I began to think that...here comes the Federal government, teaching them how to manage this type of event, after it had occurred. Kind of like when the horse is already out of the barn it is too late to close the door. I then grew concerned that this could be a hostile crowd (Fig. 2.1).

Instead what we had was a cross discipline plugged classroom all week long. Students eager to learn and share what they thought they had done well and what they thought they had not done so well. I asked them about their use of the ICS. "ICS?" they said. "We all used it. They had theirs, we had ours. Problem was we all used it individually and never went to Unified Command when we should have." My impression of what we accomplished that week together in training was more Critical Incident Stress Debriefing as much as anything. After class, one day they took us to the site of the bombing and for many of them, it was the first time they had been back since it occurred. The last day of class the Mayor presented each of us on the training team with a chunk of the building. A reminder to share anytime I am in front of students, if it can happen in Oklahoma City, it can happen here.

Figure 2.1 Chunk of marbled wall of the Alfred P. Murrah building. A token of appreciation from the mayor for the training we provided.

If an event impacts only your organization and you can manage it with in-house resources, no need for a Unified Command (UC). On the other hand, if the incident is multijurisdictional, multiagency, or requires outside assistance, such as the bombing in Oklahoma City, it is a great tool to utilize. Technically, if an organization is just providing support, you can incorporate them into a single command structure through the use of a Liaison Officer. If the supporting organization is supplying significant resources or carrying out the bulk of the current incident management tasks, asking them to join you in a UC setting may be beneficial.

UC allows different organizations to jointly coordinate, plan, and interact effectively to carry out a single set of incident objectives. Each organization maintains its authority, responsibility, and accountability. Determining who should be invited to participate with you in UC is straightforward. If they have jurisdictional or statutory authority, they must be involved. If they have something you want, invite them in.

Do not build UC like you are inviting people to a wedding. You end up inviting some people just to keep peace in the family or to balance out inviting one person with another. Participants in UC must have the authority to commit resources. If you find one of your UC participants saying they will need to call someone for an answer, they are the wrong persons. You need the person on the other end of the call. UC must be lean enough to be able to make decisions quickly. If you end up with half-a-dozen people, decision making will be slow and awkward. Organizations purely involved from a support standpoint can be managed well using the Liaison Officer.

The UC planning process includes determining:

responsibilities for incident management,
incident objectives,
limitations,
areas of agreement and disagreement between organizations,
the ICS organizational structure,
Command Center/postlocation,
unified planning process, and
unified resource management process.

Under the UC process the Incident Action Plan (IAP) is produced by the Planning Section and approved by UC. A single individual, the Operations Section Chief, directs the tactical implementation of the IAP. UC selects the Operations Section Chief based on the event needs. For example a fire in a hospital may have a UC made up by hospital and fire department representatives. During the fire suppression phase the Operations Section Chief should be from the fire department. After the fire is extinguished and area vented with fresh air, a hospital member may become Operations Section Chief to resume services for relocated patients and staff displaced from the impacted area.

Your organization would be wise to utilize preplanned events such as parades or marathons to practice UC with other organizations. The ICS is a good tool to allow organizations to work together but building relationships prior to an emergency is a proven tactic. Practicing UC in a planned event allows everyone to get more comfortable working with the tool. The advantages of using UC include:

A single set of objectives is developed for the entire event.

Communication issues can be minimized. There is no need for everyone involved in the event to talk to everyone else. Consistent with the ICS principles of Unity of Command, each UC member can talk with members of their organization. If there are joint decisions to be made, it is handled in UC and then communicated to individual organizations via their normal communication channels.

A collective approach is used to develop strategies to achieve incident objectives.

Information flow and coordination are improved between all organizations involved in the incident.

All organizations involved in the incident understand the joint priorities and restrictions or limitations.

No organization's legal authorities are neglected or compromised.

The combined efforts of all organizations are optimized as they perform their respective assignments under a single, jointly developed IAP.

Figure 2.2 MCI exercise involving EMS, fire, law enforcement, hospital, and USCG assets.

Comparison of Single Incident Commander and Unified Command
Single IC—The IC is exclusively responsible (to the extent of their authority under the ICS) for establishing the incident objectives. The IC is directly responsible for ensuring that all functional areas required are opened and staffed by competent supervisors and their activities are undertaken to accomplish the incident objectives.

UC—Individuals representing their organization must jointly determine the incident objectives and adopt one IAP for the entire event. They must work cooperatively to execute and integrate response which maximizes the use of available resources (Fig. 2.2).

What is your choice for command in this situation, single or unified and why?

If you said unified, I would agree with you. It looks like we have field triage going on so this is a larger event that could last some time and we have police, fire, and Emergency Medical Services (EMS) personnel involved. Actually, when we conducted this drill, we also flew in a physician from a local hospital by a United States Coast Guard (USCG) helicopter.

COMMAND STAFF

As part of the typical ICS organizational chart, the Command Staff includes: a Safety Officer, a Liaison Officer, a Public Information Officer (PIO) (antiquated function), and as needed a Technical

Specialist. Each of the Command Staff members reports directly to the IC or UC.

> Safety Officer—The Safety Officer monitors the incident operations and keeps the IC/UC advised on all matters relative to the safety of all personnel assigned to the organization. This includes members of the Incident Management Team as well as those deployed to carry our work effort. Incident Management Team members will sometimes work too long without appropriate rest, hydration, or nutrition which can lead to poor decision making, putting others at risk. The Safety Officer's function is one I do not think should be blended with another ICS position. I would much rather see an intermittent Safety Officer delegated buy the IC. This way they are 100% focused on evaluating the entire operation from a safety perspective and not splitting duties and/or distracted elsewhere.
>
> Liaison Officer—The Liaison Officer can act as the point of contact for supporting organizations to use in working with the IC/UC. In some cases, organizations may find it useful to deploy a Liaison Officer to another UC post, Agency Operations Center, or Emergency Operations Center if appropriate. This may be useful to improve communication and timely sharing of information and support. Support organizations communicating with the Liaison Officer must have the authority to speak for their agency/organization, after consulting with their leadership as necessary.
>
> PIO—Historically the PIO is responsible for communication with the public and media and/or other organizations with incident-related information. There are problems with organizations holding on to this antiquated view of this functional role.

Once upon a time, emergency management professionals viewed the media a necessary evil rather than a partner. Organizations either hired a dedicated PIO to keep the media at bay or assigned the task to under-utilized personnel. In some cases the media was dealt with after the fact, if at all.

The media is no longer waiting for organizations to deliver information. And the public? They are sharing disaster information amongst themselves. With the proliferation of smart devices and social media channels, anyone can publish updates on disasters in progress—even the people who are directly impacted.

There is no cushion for organizations to deal with issues outside of the public eye and a new host of communications issues have arisen: "facts" are spread without vetting, rumors run rampant, errors are not retracted, and the sheer volume of content can supersede any efforts to set the story straight or deliver timely news to impacted populations.

The solution? You can keep the PIO title if you want but the function should better reflect today's information environment. Ideally the function of communication needs to be built, cultivated, and managed well in advance of need. Throw out the idea that a "real world" reputation and "position of authority" matters online. If an agency is not active on social media channels before a disaster, key messages will not be heard once that disaster is in progress. An agency will not be sought out for information nor listened to if it has not earned that right by engaging the public well in advance and becoming known as a trusted source of timely and reliable information.

Your PIO should know services inside and out, understand the voice and personality of an agency's leadership, and, most importantly serve customers, the general public, when and how it is convenient for them. This is a critical position that serves as a conduit to the public and staff, so do not underestimate their value. Day to day a seasoned PIO must know how to deliver needed information, deal with disgruntled people, manage conversations, and know when to take conversations off-line. They can also identify areas of concern from the public and identify gaps in information, services, and needs during an event. A PIO will yield benefits to agencies before, during, and after crisis situations. Now is the time to look for the right person to fill this role.

AGENCY EXECUTIVE

Occasionally, IC is under the mistaken impression that they are the top dog. The function on an IC was discussed earlier and they do have considerable responsibility when it comes to managing a specific event. On the other hand, every IC has a boss who has influence in the priorities and the way they want an event handled. If you doubt this think back to the occupy movement active during 2011. The resolution of the occupy situations varied across the country. It was not because law enforcement did not understand their job, but due to the fact elected officials weighed in on how they wanted the occupiers handled.

Incident management roles

Incident Commander	Agency Executive
• Manages the incident • Keeps the Agency Executive informed on all important matters pertaining to the incident which have an impact on the ability to deliver services to the community	Provides the following to the Incident Commander: • Policy • Mission • Strategic direction • Authority

Figure 2.3 Chart showing difference of roles between Incident Commander and Agency Executive.

Elected officials/agency executives are expected to commit to National Incident Management System (NIMS) by supporting training and exercise participation and reinforcing the need to use the ICS in managing events. They are not expected to assume the role of IC or direct tactical operations but they do not give up their responsibilities to the community or organization either. Think of an IC as addressing a short-term issue and the agency executive looking at the long-term implications. A wise IC will use the agency executive to court those community/organizational relationships beneficial to helping them manage, resolve, and recover from an event (Fig. 2.3).

COMMAND CENTER

Your Command Center is a specific location where the hospital's incident management team have access to resources necessary to accomplish their job. Hospitals are notorious for not having much extra space and often board rooms become Command Centers by default. This can work but may not be the best arrangement. This may be a suitable place to get people together for a face to face meeting but can become crowded and noisy if your incident management team is large. With the advent of smart devices carried by most staff, activation of a virtual Command Center should be considered.

Purchasing a kit of Hospital Command Center supplies does not make people more efficient in most cases. Staff typically are most productive when operating out of their normal environment. Consider having regularly scheduled reporting face to face meetings with the entire incident management team and then let them move to a more suitable location for functions such as operations, planning, logistics, and finance/administration.

Hospitals should establish clear triggers for when their Command Center is going to be activated. It is not necessary to activate every time the ICS is used. Criteria to consider include:

scope of the event,
potential/real impact on the hospital,
available resources, and
special response needs.

Hospital Emergency Managers may want to consider tying Command Center activation to the NIMS Event Typing list. By describing what the distinct levels (1–5) look like, you will have a better chance of staff recognizing when it is appropriate to activate. This also will tie the hospital in with the community at large better for major events.

CHAPTER 3

The Operations Section

Command establishes the objectives and the Operations Section carries out the work effort to meet the objectives. This includes identifying the best tactics to accomplish the job. A smart Incident Commander can make their life much easier by assigning the right Operations Section Chief and basically staying out of the way. Command should only want to know two things from the Operations Section Chief. First, does operations need anything to accomplish the objectives? Second, they want to know when the work is done. Everything else is in the Operation Section Chief's lap.

The Operations Section Chief needs to be involved in the setting of the incident objectives to make sure they can be achieved within the time frame of the operational period. If the command has several objectives, the Operations Section Chief can keep people focused by always going back to the incident priorities built into the Incident Command System (ICS). Those priorities are:

Life safety
Incident stabilization
Property/environmental preservation
Return to normalcy

By keeping these priorities in mind the Operation Section Chief can adjust to changes in the event or resources available to avoid spreading themselves too thin. If you are working on a priority of property preservation and a life safety issue comes up, revise your work effort to address the life safety issue.

> The reader may not appreciate the value of something as simple as learning and using the four priorities but it will accomplish a couple of things. First off you will always be focused on what is important. Second your actions will be much more efficient and effective in achieving the best possible outcome. Believe it or not, it makes decision making easy, even in the face of disaster.

Figure 3.1 Motor vehicle hanging on a guardrail—fairly common occurrence. Courtesy of Patty Breun.

Let us say your driving home from work and come upon a situation as shown in Fig. 3.1.

It looks like a car has ended up hanging over a guardrail, no big deal. Priority one is life safety so we can conduct a quick patient assessment, maintain and airway or stop bleeding if needed. Priority two is incident stabilization. In this case if your initial impression is the car is not stable, you may want to secure it with some form of rope or cable just to keep the car from falling over before you do any patient assessment. In this instance, incident stabilization is also a life safety issue. Moving on to priority three which is property/environmental preservation. We could take a quick look and if we saw fuel leaking, we could toss a bucket or piece of carpet under it to help keep it contained. Damage to the car and guardrail is done, so no more work on priority three. Priority four is return to normalcy which in this case means when emergency services show up, you get back in your car and head home.

Figure 3.2 Motor vehicle debris field from a tsunami—a rare occurrence. Courtesy of Patty Breun.

Let us change up the situation just a bit and on the way home from work you see a little different situation as shown in Fig. 3.2.

Obviously, something very bad has happened. Forget panicking or not knowing what to do, go back to the priorities. Number one is life safety. When I see this, I immediately start thinking is whatever caused this done? If not, my own life could be in jeopardy and I better come up with a wise plan to extricate myself to safety. If I conclude, the situation is not ongoing then life really gets simple and I can focus on what is important. Since I am never going to get to priority two, three, or four, so just relax and do what I can to preserve lives of those I come across. My actions are defendable and more importantly I can sleep at night knowing I did what I could.

Figs. 3.1 and 3.2 are from the aftermath of the 2011 tsunami that hit Japan. If ever a situation called for some type of easy to remember and use system for prioritizing response actions, this was it.

Many National Incident Management System (NIMS) Type 5 and 4 events can be handled with one person assigned to the Operations Section. Expansion of this basic structure will vary based upon numerous considerations and factors which influence the entire operation. In some cases a functional approach makes sense. In others, there may be a need to develop an expanded Operations Section based upon geographical or jurisdictional basis. If needed, you can develop an Operations Section that is a combination. No strict rules here, just make sure that any expansion makes operational sense, has a task to accomplish, and is appropriately resourced and supervised. Some of the common elements established are discussed below.

Branches—Branches may be functional (e.g., Plant Engineering Branch), geographical (e.g., Clinic Building Branch), or both depending upon the specific circumstances of the event. Normally, branches are established when there is a span of control issue due to numerous groups and/or divisions being opened. Unless you are dealing with a major NIMS type 3 event or worse, it is unlikely you will need to open a branch very often. On the other hand, if you find your organization has a regular need for specialized support/activity during most events, it may be practical to develop a branch just for that function.

The California Emergency Medical Services Authority developed a Medical Care Branch Director position within the Hospital Incident Command System. The stated function is to provide acute and continuous care of the incident victims as well as those already in the hospital. I would suggest this only confuses the issue of who is responsible for what. If you activate the ICS for an event, the incident management team is tasked to manage the incident not patients already in the hospital. Let the hospital staff not tasked as part of the incident management organization continue to do what they should be doing. Return to normalcy is a key priority and the less people and resources you tie up in your incident management organization, the quicker you can return to normal operations.

Divisions and groups—Divisions and groups are great tools in the ICS toolbox. Divisions may be used to divide up an incident geographically such as the third Floor Division. Personnel assigned to that team would know they are going to be operating only on the third floor doing a task. The term group would be used to denote a function such as the evacuation group. By using the term group, it means personnel assigned will be doing the function of evacuation and not limited geographically. Knowing and using these terms will allow a hospital to better coordinate with local organizations in major events.

Resources—Resources assigned to the organization may be organized in three separate ways depending upon the specific requirements of the event.

Single resources—Just like it sounds, single resources means one person or piece of equipment with an operator. Use single resources if need be but keep in mind any safety issues that might be an issue. Having too many single resources assigned to one supervisor will lead to span of control problems. Think about using either a Strike Team or Task Force configuration to shrink the span of control numbers.

Strike Teams—Strike Teams are made up of identical resources. For example a Triage Strike Team would be made up of personnel having equivalent skills sets, common communications, and a designated leader as a supervisor.

Task Forces—Task Forces are combinations of resources that are designated to support a specific function and/or mission. For example a Vertical Evacuation Task Force would have sufficient trained personnel and equipment to carry out the function of vertical evacuation. Like the Strike Team they would require common communications and a designated Task Force Leader.

CHAPTER 4

The Planning Section

Different people are better suited to distinct positions within the Incident Command System (ICS) structure. When I am teaching the ICS 300 and ICS 400 program as a 5-day block of instruction, I can see these people emerge. It is a combination of experience, wisdom, and personality. Incident Management Teams (IMTs) are no different than your trauma team or decontamination team. In most healthcare emergencies, teamwork is essential to ensuring the best possible outcome (Fig. 4.1).

IMTs are made up of specific positions that work in synch to accomplish a critical mission. Each team member brings a skill set that is perishable if not maintained. Individuals tasked with being the Planning Section Chief need to have the ability to out-think the Incident Commander (IC). The IC will be focused on the here and now, and the Planning Section Chief needs to focus on what comes next. Not everyone has the right personality and way of thinking to be a great Planning Section Chief. What you will find if you get fingered as the go to Planning Section Chief, you will never work in any other position. It is that specialized and that important.

Planning is responsible for collecting, evaluating, and disseminating information to Command and members of the IMT. Planning builds the Incident Action Plan (IAP) and other status reports and documents as appropriate. We will discuss the IAP process in detail in Chapter 9, Incident Action Planning Process.

The Planning Section is made up of four primary units and technical specialists as necessary to evaluate the situation, develop planning options, and forecast requirements for the next operational period.

Resource Unit—This unit records the status of resources committed to the incident. In addition, this unit considers the effects; additional resources will have on the incident and anticipate resource needs.

Figure 4.1 Morning briefing at the 2016 Cascadia Rising earthquake exercise which included multiple military units, civilian and governmental agencies, and air assets.

Situation Unit—This unit is responsible for the collection, organization, and analysis of incident status information. They continue to analyze as the situation progresses. They can often pick up trends that others may not see.

Demobilization Unit—This unit is tasked with ensuring an orderly and efficient demobilization of all resources committed to the incident.

Documentation Unit—This unit is critical particularly when it comes to opportunities for reimbursement and protection from liability. They collect, record, and maintain all documentation relevant to the incident.

Technical Specialist—I often get asked, "How do we know we need a technical specialist?" The answer is intuitive. When the IC faces something they and their IMT have little background in, it is time to find a Technical Specialist. Admitting you do not know something and seeking help will not hurt people. Bulling ahead without a complete understanding of the situation most certainly will.

CHAPTER 5

The Logistics Section

The epitome of a Logistics Section Chief would be Corporal "Radar" O'Reilly from the M*A*S*H television series. Radar always seemed to know what was needed before it was called for and he had already located it and it was on the way. Between the Section Chiefs, operations and logistics usually work hand in hand. Operations need stuff to complete their objectives and logistics must excel at getting stuff (Fig. 5.1).

This section also provides facilities, security (for the Command Post) transportation, supplies, fuel, equipment maintenance, food service, communications and IT support, and emergency medical services for those involved in the response. Within the Logistics Section, there are some primary units, each with specific functions. As you read the description of the unit functions, you will probably be able to identify people and/or departments within your organization that manage this function on a day to day basis. This makes choosing Unit Leaders very straightforward. Letting people do what they do daily during times of disaster is the ideal solution.

Supply Unit—This unit orders, receives, stores, and processes all incident-related resources, personnel, and supplies.

Figure 5.1 Establishing a decontamination corridor can be a huge logistical burden.

Ground Support Unit—Provides all ground transportation during an incident. They are also responsible for maintaining and supplying vehicles, keeping usage data for potential reimbursement, and developing incident traffic plans.

> What type of logistical challenges would be created by a major snowstorm? Which units would you need to open and why? (Fig. 5.2).

Facilities Unit—This unit sets up, maintains, and demobilizes all facilities used in support of incident operations. They handle facility maintenance and security services required to support incident operations.

Food Unit—This unit determines food service requirements, plans menus, orders food, provides cooking facilities, cooks, servers, maintains the food service areas, manages food security, and works with the Safety Officer to ensure that hydration is available to all personnel on a constant basis.

Figure 5.2 A look down my neighborhood street after the thundersnow storm of 2011.

Communications Unit—The major responsibilities of this unit include ensuring effective communications systems are in place and operational to support the IMT and personnel involved in the incident. This may entail developing the communications plan as well as acquiring, setting up, maintaining and accounting for communication equipment assigned to the incident. Since most events and major exercises list communications as a point of trouble, early activation of this unit could make an enormous difference.

Medical Unit—Often over-looked in a hospital setting, this unit is responsible for ensuring the effective and efficient provision of medical services to incident personnel. It may seem redundant to readers that a hospital should need to designate a unit to provide medical response but it is a game changer. If there is not a team dedicated to taking care of incident personnel and someone goes down, everyone around wants to jump in and help.

If this ad hoc response is your "plan," you will have a couple of problems. First you will lose personnel that were assigned to a task designed to meet specific incident objectives. You will have more people than necessary to help the staff member who went down, so your response will neither be efficient and in some cases effective. Take the time to open this unit, develop a medical plan (ICS form 206), and communicate this plan to all personnel. Personnel assigned to this unit may also serve as Safety Officers if you are short on staff.

CHAPTER 6

The Finance/Administration Section

I mentioned earlier that operations and logistics have a natural relationship with each other. Finance/administration comes in to balance out their spend-happy life style. I have found that most organizations do not appreciate the real value of this section until they have been through a federally declared disaster and receive zero reimbursement. Not all things are reimbursable, but not opening this section and staffing it with personnel who have been educated in the Federal Emergency Management Agency (FEMA) reimbursement program guarantee you will get nothing.

> Here is something I have learned but keep in mind it is a double-edged sword. If you are the Operations or Logistics Section Chief and you think you might need something, order it early in the event before the Finance/Administration Section is established. Chances are good you will get what you want. If you wait to order an expensive item later in the event, this section will scrutinize the purchase and ask you to look at other less expensive options. Another helpful hint is keep in mind if it flies in the air, it is probably real expensive (Fig. 6.1).
>
> What I have learned is that it is sometimes nice to have someone take an objective look at what I think I need. You will also find that the personnel assigned to the Finance/Administration Section can often get services and materials faster and cheaper than going through logistics since they already have numerous contracts with vendors.
>
> The bottom line is learning to appreciate this section for what they can do to make your operation more efficient and effective. If this section is up to the speed in reporting to Command, it will make Command's job much easier when they update the agency executive on how the incident operations are progressing. Letting your agency executives be fully aware of costs ensures there will not be any ugly surprises.

Some of the functions that fall under this section are recording personnel time, maintaining vendor contracts, administering compensation and claims, and conducting an overall cost analysis of the

Figure 6.1 Emergency Medical Services (EMS) special operations class working with the US Coast Guard Jayhawk.

incident on the organization. If this section is opened they will need to coordinate closely with the Planning and Logistics Sections. If the Finance/Administration Section is not opened, consider assigning someone with this type of background to work with your Logistics Section.

The Finance/Administration Section has four primary units that have specific functions and can be opened as appropriate.

Compensation/Claims Unit—This unit is responsible for finance issues resulting from property damage, injuries, or fatalities because of the incident.

Cost Unit—This unit tracks costs, analyzes cost data, and makes recommendations regarding cost savings measures.

Procurement Unit—This unit is responsible for finance matters associated with vendor contracts.

Time Unit—This unit records and tracks the time of personnel and equipment dedicated to the incident.

CHAPTER 7

Intelligence/Investigations

This section is a relative newcomer to the world of Incident Command System (ICS). The purpose is to help with the collection, analysis, and sharing of incident-related intelligence. This function sounds like it belongs with the Planning Section and it could be placed there but the focus is slightly different. Some incidents lend themselves to requiring intelligence and investigative information collection, analysis, and appropriate dissemination.

> Case 1—The incident is such that information that leads to the detection, prevention, apprehension and prosecution of criminal activities, and the individuals involved is necessary.
>
> Case 2—The incident is such that information that leads to determination of cause, projection of spread, assessment of impact, or selection of counter-measures for a given incident (regardless of the source) is necessary. Examples might be public health events such as a cluster of cases or an unusual case or a disease outbreak.
>
> An example would be after a major flood or tsunami where you need a great deal of information regarding the impact to infrastructure and anticipated timing or likelihood of support (Fig. 7.1).

The Intelligence/Investigations Section is one section that benefits from the flexibility that comes with the ICS. This section can be placed in several locations based on incident requirements.

Within the Planning Section—This is a traditional location for the function and is appropriate if the incident does not fall into Case 1 or 2 earlier and/or there is little need for specialized information.

As a separate General Staff Section—This option may be appropriate if the incident falls into Case 1 or 2 earlier or if there is a significant criminal or epidemiological investigation occurring with multiple agencies. This may also be an excellent choice if there is a need for a

Figure 7.1 Common sight in Minamisanriku, Japan after the 2011 tsunami. Courtesy of Patty Breun.

great deal of specialized information requiring technical analysis. Examples might be a chemical, biological, or radiological event where there is a critical need for timely analysis leading to lifesaving operations. One other case would be if there is a need for classified intelligence. Keeping this function at the general staff level offers better control over the dissemination of information.

Within the Operations Section—This makes sense if there needs to be a strong link between the investigative information and the operational tactics being utilized.

Within the Command Staff—This may be appropriate for most events where there is little need for tactical information or classified intelligence but supporting agencies are providing real-time information to Command.

> Case situation. You are the on-duty supervisor at a hospital when you get notified from the local dispatch center that a cruise ship has just docked at the port. They advise you that apparently 10% of the passengers and crew are ill.

Figure 7.2 Cruise ship docked in Astoria, OR.

Question 1. Do you need to utilize the ICS? If so, will it be single or Unified Command. If unified, who should be on the Unified Command Team?

Question 2. Do you need to open the Intelligence/Investigations Section? If so, what agency should be the Section Chief from? (Fig. 7.2).

In this case, we went with Unified Command consisting of representatives from public health, hospital, and the Sheriff's office. You can certainly have different representatives, but be careful inviting too many in, have a reason. We included law enforcement from the standpoint of opening the Intelligence/Investigation Section for a joint investigation into the cause of the illness and crowd control could be a big issue.

CHAPTER 8

Area Command

You work in a hospital setting and you read the title and say, "I'll never need this." One other important lesson I have learned over time is to never say never. When we start talking NIMS Type 3 events of a complex nature or covering a wide area, I assure you that Area Command will get established. Get in to the NIMS Type 2 and 1 events and Unified Area Command will be the organization that drives the response. If your hospital is one of a dozen in an impacted area, you need to understand how to operate within an Area Command structure.

Area Command is an organization that is designed to oversee the management of multiple incidents that are being handled individually by separate ICS organizations. It may also be used to oversee the management of a very large or evolving incident being handled by multiple Incident Management Teams (IMTs).

> I had an opportunity to participate in a Unified Area Command. This was in 1996, long before the letters in NIMS had been assembled. I was working for the Oregon State Health Division as the Medical Preparedness Officer for the State. It was spring time and there had been an unusually heavy and late snow fall for this time of year. Within days of the snow event, a pineapple express warm front came through bringing heavy rainfall and warmer temperatures that started a major snow melt. The combination of snow melt and rainfall created a massive flooding problem which reached a 500-year event status. Roads, homes, and businesses were flooded, power lost, and a federal disaster was declared (Fig. 8.1).
>
> Willamette Falls seen in the photograph above normally has a 40" drop. As you can see the water level is almost flush.
>
> It was a Sunday afternoon and I was at home in Portland with the family. We had not had power for 3 days so the novelty was wearing off. The man-made lake out back was steadily rising and housing downgrade from us was starting to have water problems. I received a phone call from my boss. He told me he needed me to head out towards the coast to

Figure 8.1 Willamette Falls during 1996 flood event.

find out what was going on in a community. The State Health Division had been getting second and third hand information that there was a real flooding problem. When the State contacted the county, they were told that there were no problems.

My boss went on to explain that the State Health Officer wanted someone to establish the ground truth. I questioned why me and how this fits in with my role, after all I was a State bureaucrat now and we worked Monday to Friday. I was told I was the only one in the office with extensive field experience and had a better chance of pulling it off. No plan mind you to follow, but details ... details.

I had a couple things going for me. I had an all-wheel drive State-issued SUV, decent communication equipment, and had taught an EMT-Basic class in the community I was headed to. I was following a road that followed a small creek. As I got closer to the community I noticed the debris field was 12"–15" up into the trees along the bank. Obviously, a great deal of water had passed through here and I wondered what I would find ahead.

What I found was a devastated community. All communication was out so they had no way to ask for help. The county had mistakenly taken their silence to mean they were fine. The local volunteer fire chief was trying to run the response even though his house had been washed away in the flood. The chief had been one of my EMT students so we had a reunion of sorts and set about trying to get organized. The first law enforcement presence we saw was the sheriff from a neighboring county. Resident county deputies could not gain access due to flooded and destroyed roads.

So where does this Unified Area Command come into this story. The three of us ended up forming a Unified Area Command. The sheriff represented County law enforcement and was also statutorily responsible for emergency management. He received authority from his county's elected officials and from the local sheriff to represent his office too. The local fire chief represented the community and I was authorized by the State Health Officer to take whatever measures I deemed appropriate to avert a public health emergency. We had an IMT consisting of local, county, and state government.

The first resource request I submitted included an ICS overhead team from a private EMS provider that support wildfire operations. They brought all the ICS forms and served as the Planning Section. The next resources I requested were potable water and communication. Medical issues consisted mostly to preexisting conditions, no access to normal health care services, and loss of access to medications. We ended up opening a clinic in a church vestry staffed by a National Guard medical unit. The use of the ICS and the Unified Command process served this situation very well.

There are several factors that help determine when to consider opening an Area Command. Among them are:

1. Incident is not site specific. In the flood example, there was no one site of a problem to attempt to attack and solve.
2. The impact of the incident may not be immediately identifiable. While the flooding problem was known, the collateral issues of animal carcasses, lack of potable water and a working sewer system, and lack of access to chronic medical care and pharmaceuticals were constantly evolving.
3. The incident is geographically dispersed. The flood example involved multiple states and counties with numerous jurisdictions and organizations.
4. The incident evolves over time. In this case, the response phase of the flooding incident lasted several weeks. Once the flooding stopped, the return to normalcy phase took even longer.

Other incident examples where an Area Command may be a wise choice of an incident organizational structure are public health emergencies, earthquakes, tornados, civil disturbances, or any event where multiple IMTs are being used and the potential for competition for resources may surface.

Any major terrorist act, wildfire, or hazardous materials event will also require a coordinated intergovernmental, nongovernmental organizations (NGO), and private sector response. Large-scale coordination will take place at a higher jurisdictional level. When the incidents span multiple jurisdictions, as in the case of the flood example, a Unified Area Command should be used. This allows each jurisdiction to have appropriate representation in the Command structure.

In a major event where Area Command is implemented, your Agency Executive will be instrumental in getting your needs communicated and interests represented. If multiple hospitals are involved and competing for like resources, the adjudication of these resources will take place at the level of Area Command if established. Area Command is a good ICS toll for managing major events. Even though relatively uncommon in use, your organization should understand how to work within the structure should the need arise.

> A perfect example of a situation requiring Area Command was the 2011 earthquake/tsunami event in Japan. A very similar, if not worse, event is going to happen on the west coast of North America (Fig. 8.2).

Figure 8.2 Community shower station set up by Japanese Self Defense Force after the 2011 tsunami. Photo Courtesy of Patty Breun.

This was the public bath located in the town center of a Japanese community devastated by the event. During the day, the bath house rotated between men and women times. Many homes were without running water and the sewer system was badly damaged. Rather than try and work house by house to restore services, a decision was made at a higher level to restore service to one communal site. In addition to helping community members have access to this service, it also served as a gathering spot for people to talk and support each other. The Japan Self Defense Force is running this operation.

CHAPTER 9

Incident Action Planning Process

I will admit that when I just ran ambulance calls for a living I was not big into planning. As I moved up in responsibility overtime, I found that long-range planning became a necessary skill set. The Incident Command System (ICS) Incident Action Plan (IAP) process offers a near-foolproof method for developing plans that will work for pre-planned and spontaneous events.

RELATIONSHIP BETWEEN YOUR EMERGENCY OPERATIONS PLAN AND AN INCIDENT ACTION PLAN

Knowing how to develop an IAP does not replace the need for capstone Emergency Operations Plans (EOP) and department or unit specific procedures in the event of an emergency. Your EOP should cover strategic level issues and identify how overall command and control and resource coordination matters will be managed. To prevent over-planning and keep you National Incident Management System (NIMS) compliant, your EOP should reference that the ICS will be used to manage events.

Expecting members of your Incident Management Team (IMT) to pull out and use copies of your EOP during a fast-paced evolving event which is impacting the patient care environment is a recipe for disaster. In many facilities, there are a handful of staff who know what is contained in the EOP. In some the author of your EOP is the only one who really knows what is in it.

By using the IAP process within the ICS, your IMT can conduct real-time planning that is relevant to the incident, community impact, environmental conditions, and your organizational operating conditions. Your IMT understands the IAP well since they all helped create it. The EOP may still be used as appropriate, but it can be utilized when the timing is appropriate.

UNDERSTAND THE SITUATION

The first step in planning for anything is taking the time before hand to understand as much about the situation as possible. If your facility sits close to an industrial area where hazardous chemicals are used in manufacturing, you should know as much as possible about those materials. If your facility is near a river, stream, or area influenced by the tides and weather, you better understand flood mitigation and response strategies and concepts. Active shooter, or acts of violence are becoming more common place so take the time to learn the facts about such events. I say facts because there is a good deal of misinformation and notional boiler-plate response plans that get touted as doctrine.

> When I was hired to be the Medical Preparedness Officer for the State of Oregon, one of my primary responsibilities was to conduct and hazard assessment and gap analysis if we ever had a problem at Umatilla Army Depot. The depot was home to about 4000 tons of GB, VX, and sulfur mustard. Military chemical warfare agents left over since WWII. No chemical weapons were used against troops in WWII but at the close of WWI every other shell was a chemical munition so we had built up a huge stockpile, assuming the next war would be a chemical one. The introduction of nuclear weapons was assumed to be a sufficient deterrent against another country using chemical munitions. This resulted in the munitions being deemed obsolete and scheduled for destruction. Keep in mind most of these were decades old and had never been intended to last this long so leaking and instability became issues.
>
> So, what does this have to do with anything you ask? At this point in my career, I understood public safety organizations quite well and the normal disaster response process. What I did not understand were military chemical warfare agents. I told my boss I needed to go to school to learn about these products to plan effectively. Who else better to teach me about military warfare agents than the military. In short order, I was off to Aberdeen Proving Grounds to attend the Medical Management of Chemical and Biologic Casualties course offered by the United States Army Medical Research Institute of Chemical Defense (USAMRICD).
>
> It was at this course that I met Dr. Fred Sidell. Dr. Sidell literally wrote the book on medical management of chemical casualties. Dr. Sidell and I just hit it off and together we wrote the medical curriculum for use in training healthcare providers in Oregon. One year later, I would be invited to go through USAMRICD's validation process to become an instructor in these topics and get to teach alongside Dr. Sidell and his peers both in country and internationally (Fig. 9.1).

Figure 9.1 Doing a little shopping in Naha, Okinawa.

In addition, I was put on a National-Level Task Force to evaluate Personal Protective Equipment (PPE) and detection equipment that could be used by civilians when up against military agents. The point of sharing this experience is that learning about the military chemical agents let me plan much more effectively. In this case, it was important to know what you need to worry about when dealing with military weapons but just as critical was knowing what not to worry about.

The best place to start understanding the situation is with your facility Hazard Vulnerability Assessment (HVA). Take the time to develop short summaries of the threats and hazards and give case examples for personnel to review. I know we all like to think we are special but the truth is some organization somewhere has dealt with a similar event. Learn what you can from After Action Reports. If you do not take the time to understand the situation before an event happens, you are on a steep learning curve so open that technical specialist box and staff it with a knowledgably source you trust. Do not be afraid to go outside your organization to staff it if that results in the best person for the function.

DETERMINING YOUR OPERATIONAL PERIOD

I think this is one of the most important steps in developing your IAP, even more than the objectives themselves. Before you can say what you want to accomplish, you need to determine under what time period you want your objectives to be accomplished. This helps you to build realistic objectives and keep them achievable.

Most federal training doctrine references a 12-h operational period. That may be useful in some cases and it is an effective way to increase total personnel available for your operations if you run a three-shift staffing model. By making the operational period 12 h, you eliminate one whole 8-h shift and can divide them between the two 12-h teams. I do not think this is always a good option for healthcare organizations since there are so many specialist positions. If you just need warm bodies, a 12-h shift may be fine. On the other hand, if you need key positions filled with limited staffing, you may need to be more creative.

There are no rules when it comes to setting an operational period however you should consider incident complexity, scope, and dynamics. Generally speaking, I recommend your initial operational period be kept fairly short (2–4 h), particularly if the event is quickly evolving or of higher risk. It is natural to try and write more objectives the longer you make the operational period. By keeping your first one or two short, it forces you to be very deliberate and focused in setting the objectives.

Another common-sense approach is to set your first operational period to coincide with wherever you are in your organization's schedule. For example, if you run three 8-h shifts and you activate your team 3 h into your shift, your first operational period should be no longer than 5 h. The idea is that you do not change people's schedules arbitrarily, keep disruptions to services to a minimum and (your finance folks will love this), you do not get into any overtime issues.

If you find yourself in a protracted event such as a flood or snow storm, a 12-h operational period may make some sense since the event is slower moving and unlikely to change dramatically. Another option is to go with a day shift and a night shift. Bottom line set your operational periods based on operational needs and adjust them as you move through and event and into recovery.

ESTABLISH INCIDENT OBJECTIVES AND STRATEGY

Once you have nailed down your operational period, it is time to establish your incident objectives and evaluate potential strategies. The ICS comes with a simple prioritization method for setting objectives. The concept is to ensure you focus on what is most important. It is very easy in the face of chaos and confusion to have trouble wrapping your head around everything that needs doing.

> Hospitals are often challenged when it comes to understanding the value of using of the Hospital Incident Command System (HICS) to manage events since they already have hierarchies with reporting chains. I will use the example of an Myocardial Infarction (MI) patient presenting to their Emergency Department (ED) to show the concepts of the HICS are firmly in use in the hospital setting when it comes to managing critical events. The ICS uses one set of objectives and they are consistently addressed in order of
>
> Life safety
> Incident stabilization
> Property/environmental conservation
> Return to normalcy
>
> Now follow my example of a middle-aged man patient presenting to your busy ED complaining of crushing substernal chest pain. He is pale, diaphoretic and says he "feels like he's going to die." At this point, someone steps up to take charge (the Incident Commander) and they conduct a more focused assessment to determine the extent of the problem. In the ICS we call this assessing the scene to help determine the immediate issues and formulate our initial action plan.
>
> With our MI patient, we must ask ourselves do we have a life safety issue? The answer in this case is we obviously do therefore we need to immediately assign resources to manage this patent or risk the patient's condition deteriorating further. We do not ignore patients already in the ED and we do not assign more resources than necessary. One of the key benefits of using the ICS is our ability to assign and manage resources.
>
> Next, we need to stabilize the patient by placing them on oxygen, starting a line, taking vitals, hooking up the Electrocardiogram (EKG), and managing pain. This sounds an awful lot like "incident stabilization in ICS terms, doesn't it?"
>
> We are very concerned about trying to conserve heart muscle in this case. The more heart muscle we can conserve at this point and reoxygenate, the better our patient's prognosis to survive the immediate event. I had say this is identical to the ICS task of property conservation.

> Lastly the patient is managed and a care plan initiated so upon discharge the patient returns to as near as possible their preexisting health condition, if not better. In ICS terms, we ensure a return to normalcy as soon as possible. Not just for the victims in the event but also the resources we had assigned to manage the event. When the ED personnel hand off the patient's care to the unit, they return to their work in the ED. It is all about resource management.

Objectives tell your Operations Section Chief what you as an Incident Commander (IC) want to see accomplished over a set operational period. As the Operations Section Chief, you need to identify the best strategy to accomplish the objectives. Once you select a strategy you identify the tactics best suited to carry out the strategy. Lastly implementing tactics leads to tasks so you need to select people and resources best suited for the mission.

If you start to mentally adopt this flow of objectives need strategies, strategies need tactics and tactics require tasks it is amazing how easy it becomes to manage events even outside your area of expertise.

Personnel carrying out tasks will need to have prerequisite knowledge, skills and ability (KSA) requirements. Before a major event happens is the time to identify what KSA are necessary to successfully execute the mission safely and effectively. Proper use of the ICS planning process will allow you to document these training needs and help justify your budget requests to obtain the training required.

Your objectives should conform to legal and statutory obligations and be consistent with the overall intent of the policy maker(s). I always encourage ICs to have a frank discussion with their agency executive prior to getting to far ahead in their planning process.

> An example of the role agency executives play in setting operational direction may be found in reviewing the Occupy Wall Street movement of 2011. The protests sprung up in cities across the United States but the resolution of each occupy protest varied widely. The reason they varied was not because law enforcement was confused what their response should be when people break the law. The difference was born out of the direction the political leaders in each city chose to authorize. Heavy-

> handed approaches would be viewed as too aggressive for some. A hands-off approach would be viewed as too weak by others.
> Politicians and agency executives sometimes have a perspective that differs from those on the front lines. It is good to know what action will be supported before you get to far into your planning. It is not a terrible thing or a good thing, it is just one of the ground rules you need to understand as an IC and/or Planning Section Chief. If you are in the role of Planning Section Chief and have any doubts what will be supported, get with your IC and make sure you have the backing of the policy makers.

The purpose of the IAP is to provide all incident supervisory personnel with direction based on the objectives set for that operational period. Keeps everyone on the same sheet of music, understand their line of supervision, and helps avoid duplication of effort or something slipping through the cracks.

One of the common techniques for developing good incident objectives is using the mnemonic S.M.A.R.T. This stands for Specific, Measurable, Achievable, Realistic, and Timely. It is not easy to write objectives that hit all these points but it is well worth the effort. It takes practice so once again a great reason to war-game your way through your HVA.

> Let us use a real-life disaster to review the concept of why it is important to use the S.M.A.R.T. process to build objectives. Fig. 9.2 is a hospital on the Japanese coast that was hit by the 2011 tsunami. Even though they were supposed to be in a safe zone, you can see the water mark well up the first-floor wall. Now that you know what the conditions are, let us look at some objectives.
> Let us say the water has receded and the IC asks you to "go make sure everyone is OK." How well does this fit the S.M.A.R.T. criteria. Is it specific enough for you to know what you are supposed to do? Is it measurable enough you know how to put people into OK and not OK categories? Is it achievable? Can you actually carry out the task or do you need help? How many people are supposed to be helping you? Is it realistic for one person to search the entire hospital and determine "is everyone OK?." Last is it timely and speaking of time, how much time should this task take you?
> It is obvious the objective of making sure everyone is OK is not likely to work or yield any useful information to the IC in managing this event.

We all understand the IC was concerned for everyone and needed some type of head count and picture of how bad the situation was. The problem is when Command fails to quickly set the course of action for personnel with clearly articulated actions, people will take it upon themselves to act. It is often called free-lancing and viewed as a terrible thing. It is a terrible thing but it reflects on Command and not the people who are taking action. If Command is slow or indecisive, you cannot blame others from acting.

Let us try a different approach to our scenario. How about this time you take Command and say, "Grab four people and send them in teams of two to each floor to get a quick count if any staff are hurt or any patients on life-support equipment are in immediate trouble. Report back to me here in 10 minutes." It is specific in limiting what you are asking them to do. It is measurable and you know it is completed when they report back. It should be achievable if the team stays focused and gets up and back or has communication equipment they can contact you with. You did not send one person to each floor which would be faster because this is not a normal environment and you want to enforce the buddy system at this point. It seems realistic that in a facility this size we can get it done. As for being timely it will take no more than 10 min and will give you a good picture of life safety problems. Remember priority number one is life safety. You can worry about priority number two, incident stabilization after you are satisfied that life safety is covered.

Figure 9.2 Motoyoshi Hospital where patients and staff were forced to vertically evacuate and then resume operations after the 2011 tsunami. The waterline can be seen on the building. Courtesy of Patty Breun.

Instituting the ICS by itself will not bring order to chaos but using the concepts and principles of ICS, and developing S.M.A.R.T. objectives can help you start getting things organized. I equate managing a major event to the task of eating an elephant. You need to take it a bite at a time. By keeping your objectives S.M.A.R.T., you and your IMT will show progress. Little victories at this point will rally those around you and build momentum to keep the wins coming.

The best way to get better at developing objectives is to practice regularly. Pick one event from your HVA per month and do a simple scenario driven drill to write five S.M.A.R.T. objectives. Overtime you will find it becomes much easier and you start thinking in the order of incident priorities. You will impress your supervisor with how organized you always are under pressure.

DEVELOP THE PLAN

The next task in the planning process is to develop a written plan. A quick verbal plan like we used to check on the welfare of staff and patients is great to get our response started. Now we need to start working on a plan to get us through that first operational period. I cannot tell you how many times I have heard the comment "we don't have time to plan" or "we're too busy responding to plan right now." All that tells me is your scene is out of control, resources are being poorly utilized, duplication of effort is occurring, and something is going to fall through the cracks. If no one gets hurt, it is a miracle.

Planning is the key to managing events by getting ahead of them. I can tell how experienced an IC is by how quickly they open the Planning Section and task them to start working on a plan for the next operational period. If the event in front of you is either unusual or going to last longer than 2 h, open the Planning Section and put together a written plan. Length of the plan is not important; the activity of planning is.

PREPARE AND DISSEMINATE THE PLAN

This next phase involves preparing the plan in a format that is appropriate for the complexity of the event. Often the first plan the IC comes up with is verbal, like our example of searching the hospital posttsunami. The library of initial IAPs in Chapter 10, Hospital Incident Command System Initial Incident Action Plan Library for Common Events, is good first written plans but only for the initial operational period typically associated with life safety priorities. They do not replace the need to conduct more formal planning if the event calls for it. We will look at this more formal planning process in more depth under the Planning P.

EXECUTE, EVALUATE, AND REVISE THE PLAN

The ICS planning process includes the requirement to implement the authorized plan and evaluate if the activities being undertaken are meeting the objectives. The general staff should regularly compare the planned progress with actual progress. If deviations are occurring or objectives are not being met a review of causes should be considered. In my experience the most common reasons for objectives not being met are

Your objectives were not S.M.A.R.T.

The solution is to revise your objectives. Take a little time and have your Operations Section Chief help you.

Operations do not have the appropriate personnel (in numbers or KSA lacking).

I only wants to know two things from my Operations Section Chief if I am the IC. First if they need anything and second let me know when the objectives are complete. There needs to be a good dialog between the IC and the Operations Section Chief to make sure they have what they need to be successful.

Operations do not have the appropriate equipment for the tasks.

Power equipment and equipment with wheels always make a task go easier. Make sure operations have what they need so they are not wearing out personnel assigned. Asking staff to work without the proper equipment will lead to injuries.

Working conditions have changed and/or are worse than anticipated.

It is easy for members of your IMT to forget that the conditions personnel are operating under may differ greatly from the conditions they are experiencing in the command center.

Micromanaging is occurring.

The IC needs to take a good look and make sure they are not the guilty party before getting a meeting with your Operations Section Chief. One of the biggest reasons why responsibility is not delegated is a loss of confidence in others to complete a task. The solution here is training. Another reason micromanagement may occur is you have the wrong individual in a supervisory position within the ICS.

This evaluation process is critical since it will form the first step of the formal IAP development process discussed next. If the issues are serious you can revise your current plan or use the information in developing the IAP for the subsequent operational period.

THE PLANNING P

The United States Coast Guard (USCG) uses the ICS on a regular basis and the United States Coast Guard Incident Management Handbook (USCG IMH) app is a must download for anyone on your IMT. It has a tremendous amount of easy to follow information and you can access it anywhere you can get a signal on your smart device. The USCG developed the Planning P as shown in Fig. 9.3 and it is a great tool which I suggest you to adopt and use.

Let us take a walk around this Planning P and get a feel for how it works and who does what. First off if you notice the stem of the P is the initial on-going response. To say you do not have time to plan because you are too busy responding only tells me you do not understand the ICS. Response is occurring and problems are being addressed based upon the IC's verbal plan or initial IAP. If we do not make the effort to get into the planning process, you will spend all your time reacting to events instead of managing the event.

Recognizing that response is underway, let us look at the first step in the planning process.

IC/Unified Command (UC) develop/update objectives meeting—Just like it sounds this is where objectives for the next operational period

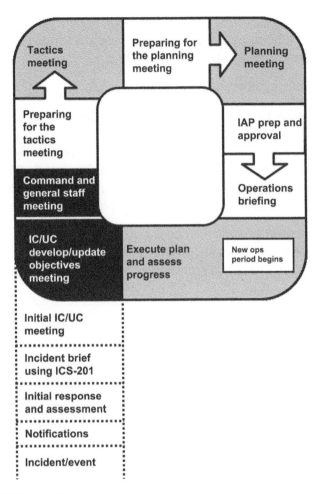

Figure 9.3 USCG Planning "P."

are developed. Since response is taking place, the IC/UC have an opportunity to examine where they currently are and where they want to be after the next operational period. They will work through the incident priorities we discussed above always making sure life safety is addressed followed by incident stabilization, property/environmental preservation, and return to normalcy. It is very possible that concurrent objectives will be developed that address several of these priorities in one operational period. I always encourage organizations to start looking at return to normalcy well before you get to that point. You may need to open a long-range Planning Unit to handle the task but the sooner the better.

Command and general staff meeting—I like to think of this as a reality check of what the IC has come up with. A good IC will solicit input and always make sure their Operations Section Chief is onboard and logistics can support their needs.

Preparing for the tactics meeting—This would be led by the Operations Section Chief with whomever they think they need to come up with the best strategies and tactics to accomplish the objectives.

Preparing for the planning meeting—This is led by the Planning Section Chief and normally has operations and logistics present. Finance can be consulted if need be. The written plan is the outcome of this meeting.

IAP prep. and approval—The written IAP is presented to Command and they sign it when they approve the document. This is a simple act but sends a powerful message. Your facility EOP states the HICS will be used to manage events and should be signed by your agency executive. People not working under the IAP may question who is doing what and why, but the fact is that this document is what must be followed by those operating under it.

Operations briefing—At this meeting all supervisors will hear from Command and each Section Supervisor as appropriate about what will occur in the next operational period. Each Branch, Group, Division, Task Force, or Strike Team supervisor will get their individual assignments which they can then take to their teams, brief them, and get prepared to go to work.

New operations period begins—At the predetermined time the elements activated under operations for the new operational period start their work effort. No longer are you reacting to events, you are out in front and simply working to complete your assignment.

Execute plan and assess progress—We have come a full circle and back to assessing how well we are meeting our current objectives. This is when we start the process to develop objectives for the next operational period.

FORMS

For anyone who has taken a boiler-plate ICS-300 course, this is where your eyes start to glaze over with the mass amount of forms. There is a need for forms but there is no law that says you must use the exact same forms that the federal emergency management agency (FEMA) text uses. You need to take the time to customize the IAP forms you will use most often so they work for your facility. If you use a commercial kit of forms or print off material from FEMA or california emergency medical services authority (CAEMS), your staff will spend more time to how to fill out the form than they do managing the event.

It is a real shame when I see organization's struggle with the HICS. Most often it is because they try and make the way they do day to day business conform to the FEMA or HICS forms. Forget that concept and make the forms conform to the way you are comfortable. For example the ICS 213 form is the message form. Rather than use the FEMA form to try and get your staff to send messages during an emergency, figure out what you use daily, put the number 213 at the top and bingo. Problem solved.

The form I use in Chapter 10, Hospital Incident Command System Initial Incident Action Plan Library for Common Events, for the Initial Action Plans is probably the only form your organization will need for most events. If you do get into needing the more formal IAP development process there are some forms I think you should know. Again the key is to customize each of these so they make sense to you and the way you normally conduct business. It is important to retain the correct form number for the document so it can be shared with other organizations involved or in a Unified Command setting.

ICS-201—incident briefing—This form has four sections but does not need to be four pages. It is designed to capture the vital incident information prior to starting the formal IAP development process. This form allows for a concise yet complete briefing during a transition from the current IC to the incoming IC. The form is not included as part of the formal IAP to be developed. If the event can be resolved before needing another operational period, this may be all the documents you need.

ICS-202—incident objectives—This form serves as the first page of your written IAP. It includes incident information, a listing of the objectives for the operational period, weather and safety information, and a table of contents for the IAP.

ICS-203—organizational assignment list—This form is typically the second page of an IAP and it provides a full listing of management and supervisory personnel and their positions for the operational period.

ICS-204—assignment list—There may be more than one 204 with the IAP based upon the organizational structure of the Operations Section for the operational period. Each division/group and potentially Task Force or Strike Team will have their own 204. The form will list the supervisors for each and the specific assigned resources with the leader name and number of personnel assigned to them. This form also describes in detail what their mission will be during the operational period in support of meeting the overall objectives. Special instructions may be included and components of the ICS-205 that are applicable.

ICS-205—incident communications plan—The 205 is designed to capture all communication assignments including any resources working under a 204.

ICS-206—medical plan—This form is to capture the procedures to be followed in case someone working under the IAP has a medical emergency. This form is not used nearly as often as it should be and chaos will reign if one of your team is injured and personnel are unsure how to address the issue. Unplanned response typically includes more resources than required and major disruption to incident operations. Just because you work in a healthcare environment do not assume this issue is taken care of.

ICS-209—incident status summary—This form collects incident decision support information and is a primary mechanism for reporting to support organizations, coordination entities, and the agency executive.

ICS-211—incident check-in list—This form is used to document the check-in process. Information from this form is typically reported to the Resources Unit if activated or to the Logistic Section Chief.

ICS-215—operational planning worksheet—The 215 is used during the planning meeting to develop tactical assignments and identify resource needs to support the selected strategies in meeting the incident objectives.

ICS-215A—safety and risk analysis—This form is used for evaluating activities against potential health or safety issues. The form includes mitigation strategies including equipment or supply needs. This is normally completed by the Safety Officer with input from appropriate sources including technical experts.

KEYS TO SUCCESS

Proactively planning is the only way to get in front of an event to manage it. In my experience, you get one chance to get a dynamic emergency event organized. By using the IAP development process your chance of success increases dramatically. The key factors in making a great IAP include:

First determine your operational period

Prioritize and write S.M.A.R.T. objectives in the order of

Life safety
Incident stabilization
Property/environmental conservation
Return to normalcy
Build your organizational chart to meet your objectives
Staff the functional positions with the best qualified person available

The planning process may seem foreign to you at first but the more you complete IAPs the easier and faster you will become at completing them. Having a written plan is a perfect way to make sure everyone is using the same sheet of music. It also prevents confusion in your personnel. I have been on several events where it was obvious someone had a plan, but they were not sharing it with those of us working.

An IAP indicates exactly who is assigned and working the event. By using the ICS, you will be closer to achieving incident priority number four, return to normalcy. People not working under the IAP should be going about their normal duties. As objectives are met under the IAP, demobilization of assets occurs which gets personnel and resources back into day to day operations. Once you gain confidence in the ICS and the IAP process you will find yourself using it more and more due to the efficiency it affords and effectiveness in organizing personnel and resources to accomplish specific objectives.

CHAPTER 10

Hospital Incident Command System Initial Incident Action Plan Library for Common Events

By now the reader should understand the Incident Action Plan (IAP) development process. The library of Initial IAPs contained in this chapter is not to replace your need to develop IAPs during events. The intent of this chapter is to jump start the incident management process. You will notice that the operational period length for each of these is 1–2 h with an average of six Incident Command System (ICS) functional positions staffed. This may not sound like much time but it buys your Planning Section Chief 60–120 min to get the IAP for the next operational period done. It is more about getting an organizational structure in place to manage the event than it is to solve all problems.

There are several threats and hazards in your local Hazard Vulnerability Analysis that you will not find in my library of Initial IAPs. Some events are slow enough developing such as hurricanes and wildfires that you will have ample time to develop an IAP. Others are complex enough that they demand their own annex development in your Emergency Operations Plan. Examples could be facility evacuations or direct attacks on the facility.

The Incident Response Guides offered by the California Emergency Medical Services Authority are a great tool to use when developing your own IAPs. I would suggest you to make separate piles for objectives you want to accomplish in the first 1–2 h, 2–6 h, and 6–12 h from the suggestions they offer. This would keep your Incident Management Team (IMT) members from being over-whelmed by the sheer number of considerations if you use them right out of the box during an event.

I developed and taught courses for the Department of Homeland Security's Center for Domestic Preparedness for about 6 years. One of the courses was a planning course for a weapons of mass destruction event. The first 2 days of the course taught students how to assemble a planning team and build an IAP for a planned community event based upon current threat intelligence. Day 3 of the course taught students went through a day long table top as the event unfolds (Fig. 10.1).

What I always found fascinating on Day 3 was the student's failure to use the written IAP they spent the 2 previous days developing. Pandemonium would rein and they would try and manage the event by the seat of their pants. Occasionally, I would point out the copies of their IAP sitting in front of them. Within the IAP they had addressed most of the issues now in front of them during the exercise. They would all admit that it would have been much easier and less stressful if they had used it from the beginning.

Figure 10.1 Posing with writing team at the Center for Domestic Preparedness.

Sit down with members of your IMT and go through the Initial IAPs I have included. Discuss them and refine to make them better for your organization. Keep in mind the third shift may not have the ideal staffing numbers so keep your organizational charts realistic. Calling in additional IMT members becomes a key activity if you anticipate the event being of any significant duration.

INITIAL INCIDENT ACTION PLAN FORM—ACTIVE SHOOTER EVENT

1. Incident Name	2. Initial Operational Period – 1 hour

3. Situation Summary — HICS 201 —

A recently discharged patient returns to the hospital with a handgun and shoots the first three people they encounter, including one security officer. Staff follow their active shooter training and the assailant while searching for more targets is confronted by local law enforcement and shot when they fail to follow the officer's commands.

4. Current Incident Management Team (fill in additional positions as appropriate) — HICS 201, 203 —

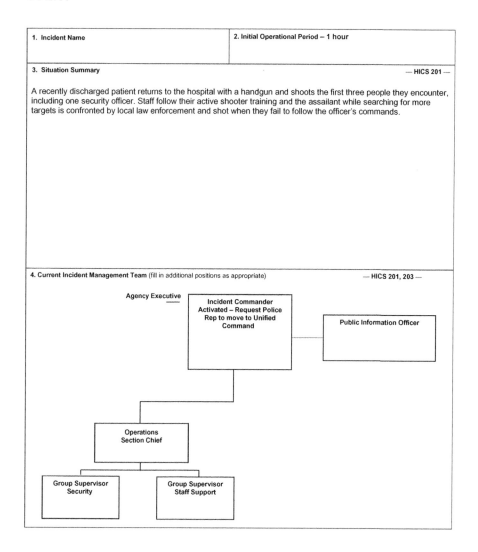

5. Health and Safety Briefing Identify potential incident health and safety hazards and develop necessary measures (remove hazard, provide personal protective equipment, warn people of the hazard) to protect responders from those hazards.	
Hazard: Emotional and psychological trauma for staff/families	Mitigation Strategy: Offer CISD, quiet location and have staff contact families/close friends
Hazard:	Mitigation Strategy:
Hazard:	Mitigation Strategy:

6. SMART Incident Objectives – Specific – Measurable – Achievable – Realistic – Timely	Assigned To:
LIFE SAFETY 1:	
LIFE SAFETY 2:	
LIFE SAFETY 3:	
INCIDENT STABILIZATION 1: PROVIDE SITUATION UPDATE TO STAFF AND STAKEHOLDERS	PIO
INCIDENT STABILIZATION 2: PROVIDE ONE ON ONE SUPPORT AS NEEDED FOR STAFF	
INCIDENT STABILIZATION 3: COORDINATE ENHANCED SECURITY AND CRIME SCENE PROTECTION WITH LAW ENFORCEMENT	SECURITY GROUP
PROPERTY/ENVIORNMENTAL CONSERVATION 1:	
RETURN TO NORMALCY 1: AUGMENT/REPLACE STAFF TRAUMATIZED BY THE EVENT BASED UPON ASSESSMENT OF STAFF AND SERVICE DELIVERY NEEDS	STAFF SUPPORT GRP

RESOURCES ASSIGNED THIS OPERATIONAL PERIOD			
DIVISION/GROUP NAME	SUPERVISOR	STAFF #s	COMMUNICATIONS
Security Group			
Staff Support Group			

7. Prepared by PRINT NAME: _____ SIGNATURE: _____

DATE/TIME: _____ FACILITY: _____

INITIAL INCIDENT ACTION PLAN FORM—EARTHQUAKE EVENT

1. Incident Name	2. Initial Operational Period - **2 hours**

3. Situation Summary — HICS 201 —

While your facility may be an area which historically has been seismically active, no one in the history of the organization ever felt such strong shaking that was just experienced. It started out as mild shaking and progressed to a wider wave length motion and continued for 60-90 seconds. Lose items on shelves and desks hit the floor, computer monitors included. In patient rooms, some light fixtures and wall-mounted TVs fall. Large equipment not bolted to the floor moves to the extent of power supply lines and rolling carts are up-ended.

4. Current Incident Management Team (fill in additional positions as appropriate) — HICS 201, 203 —

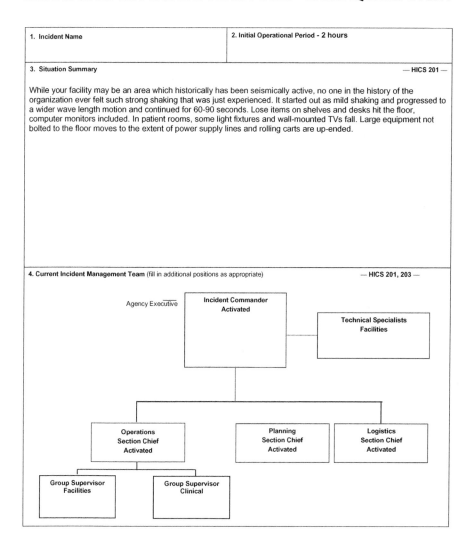

5. Health and Safety Briefing Identify potential incident health and safety hazards and develop necessary measures (remove hazard, provide personal protective equipment, warn people of the hazard) to protect responders from those hazards.

Hazard: Broken glass, metal edges, trip hazards	Mitigation Strategy: Announce hazards, avoid unnecessary movement, hard-soled shoes if available and leather gloves if handling debris.
Hazard: Downed electric lines and gas line ruptures outdoors	Mitigation Strategy: Before people exit buildings survey the outside for hazards and isolate dangerous zones.
Hazard: Unsafe areas in building do to failure of construction	Mitigation Strategy: Barricade off areas found in quick survey

6. SMART Incident Objectives – Specific – Measurable – Achievable – Realistic – Timely	Assigned To:
LIFE SAFETY 1: DETERMINE IF LIFE SAVING PATIENT CARE HAS BEEN IMPACTED AND OFFER SUPPORT	CLINICAL GROUP
LIFE SAFETY 2: CONDUCT QUICK STRUCTURAL INTEGRITY SURVEY AND CLOSE OFF UNSAFE AREAS	FACILITIES GROUP
LIFE SAFETY 3: EVALUATE NEED FOR HORIZONTAL, VERTICLE OR FACILITY EVACUATION DUE TO THREAT	COMMAND/OPS/EXEC
INCIDENT STABILIZATION 1: IDENTIFY HAZARDS AND MARK OR RESTRICT ACCESS	FACILITIES GROUP
INCIDENT STABILIZATION 2: NOTIFY STAFF AND STAKEHOLDERS OF CURRENT STATUS OF OPERATIONS	COMMAND/PIO
INCIDENT STABILIZATION 3: INCREASE ED STAFFING ANTICIPATING INFLUX FROM COMMUNITY	CLINICAL GROUP
PROPERTY/ENVIORNMENTAL CONSERVATION 1: DETERMINE STATUS OF UTILITIES AND SHUT OFF PRN	FACILITIES GROUP
RETURN TO NORMALCY 1:	

RESOURCES ASSIGNED THIS OPERATIONAL PERIOD			
DIVISION/GROUP NAME	SUPERVISOR	STAFF #s	COMMUNICATIONS
Facilities Group			
Clinical Group			

7. Prepared by   PRINT NAME: _____   SIGNATURE: _____

DATE/TIME: _____   FACILITY: _____

Hospital Incident Command System Initial Incident Action Plan Library for Common Events

INITIAL INCIDENT ACTION PLAN FORM—FLOODING EVENT

1. Incident Name	2. Initial Operational Period – 2 hours

3. Situation Summary — HICS 201 —

Flashflood - A thunderstorm with an associated heavy downpour on already saturated ground has caused flashing flooding in a nearby creek and along plugged storm drains. Large puddles have formed in the parking lot and small streams have formed which have the potential to impact lower entrances.

Slow Rising Flood - Weather experts describe the flooding as a 100-year event. A combination of increased urbanization and concrete coupled with the historic rainfall has resulted in rising water closer to your facility than ever seen before. Weather forecast indicate several more days of rain are certain and water levels will continue to rise.

This Initial IAP may be utilized in the event of either situation. The only difference is the Planning Section will have the task of developing an IAP for the contingency of either vertical evacuation or complete facility evacuation.

4. Current Incident Management Team (fill in additional positions as appropriate) — HICS 201, 203 —

- Agency Executive
- Incident Commander — Activated
  - Technical Specialists — Weather and/or hydrologist specialists
  - Operations Section Chief — Activated
    - Group Supervisor — Ingress and Egress
    - Division Supervisor — Flood Fight
  - Planning Section Chief — Activated
    - Group Supervisor — Evacuation
  - Logistics Section Chief — Activated

5. Health and Safety Briefing Identify potential incident health and safety hazards and develop necessary measures (remove hazard, provide personal protective equipment, warn people of the hazard) to protect responders from those hazards.	
Hazard: Flood waters in low lying areas	Mitigation Strategy: Staff reminders and signage at locations
Hazard: Driving hazards	Mitigation Strategy: Communicate with staff on turn around-don't drown
Hazard: Electric shock	Mitigation Strategy: Shut off power, signage barricade locations

6. SMART Incident Objectives – Specific – Measurable – Achievable – Realistic – Timely	Assigned To:
LIFE SAFETY 1: COMMUNICATE TURN AROUND-DON'T DROWN TO ALL STAFF	COMMNAND/PIO
LIFE SAFETY 2: FLOOD WATERS AND LOW LEVEL ELECTRICAL OUTLETS- SHOCK HAZARD	FLOOD FIGHT DIV
LIFE SAFETY 3: AS APPROPRIATE PLAN FOR EVACUATION	PLANNING & EVAC
INCIDENT STABILIZATION 1: SIGNAGE, FLAGGING AND/OR BARRICADES FOR FLOODED ROADS	INGRESS/EGRESS GROUP
INCIDENT STABILIZATION 2:	
INCIDENT STABILIZATION 3:	
PROPERTY/ENVIORNMENTAL CONSERVATION 1: SANDBAG LOW LYING AT RISK LOCATIONS TO DIVERT WATER	FLOOD FIGHT DIV
RETURN TO NORMALCY 1:	

RESOURCES ASSIGNED THIS OPERATIONAL PERIOD			
DIVISION/GROUP NAME	SUPERVISOR	STAFF #s	COMMUNICATIONS
Ingress/Egress Group			
Flood fight Division			
Evacuation Group			

7. **Prepared by**   PRINT NAME: _____   SIGNATURE: _____

DATE/TIME: _____   FACILITY: _____

INITIAL INCIDENT ACTION PLAN FORM—ICE STORM

1. Incident Name	2. Initial Operational Period – 2 hours

3. Situation Summary — HICS 201 —

Your area has been experiencing sever colder weather and neighboring jurisdictions have had numerous issues with the ice storm bringing down trees, powerlines and making driving conditions extremely hazardous. Your facility has already seen suppliers contacting your organization to say they are anticipating delays in making normal deliveries until the storm breaks or road conditions improve.

4. Current Incident Management Team (fill in additional positions as appropriate) — HICS 201, 203 —

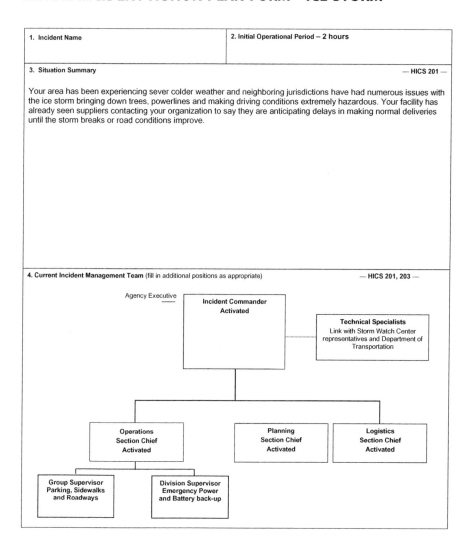

5. Health and Safety Briefing Identify potential incident health and safety hazards and develop necessary measures (remove hazard, provide personal protective equipment, warn people of the hazard) to protect responders from those hazards.	
Hazard: Slips and falls outdoors	Mitigation Strategy: Remind staff of danger and treat walkways, parking
Hazard: Exposure to cold	Mitigation Strategy: Wear warm clothing during outdoor work/movement
Hazard: Driving during storm	Mitigation Strategy: Evaluate options to house staff or call in early before weather eliminates option

6. SMART Incident Objectives – Specific – Measurable – Achievable – Realistic – Timely	Assigned To:
LIFE SAFETY 1: ADJUST STAFF SHIFTS TO KEEP PERSONNEL OFF THE ROADWAYS IF POSSIBLE	AGENCY EXEC
LIFE SAFETY 2: EVALUATE ABILITY TO OPERATE ALL LIFE-SUSTAINING PATIENT CARE EQUIPMENT IN THE EVENT OF POWER LOSS	EMERG. POWER DIV
LIFE SAFETY 3: EVALUATE POSTPONMENT OF ELECTIVE AND OUTPATIENT SERVICES UNTIL ROADS IMPROVE	AGENCY EXEC
INCIDENT STABILIZATION 1: DETERMINE EMERGENCY BATTERY NEEDS	
INCIDENT STABILIZATION 2: DETERMINE BURN RATES OF CRITICAL SUPPLIES AND DELAYED DELIVERIES	LOGISTICS
INCIDENT STABILIZATION 3:	
PROPERTY/ENVIORNMENTAL CONSERVATION 1:	
RETURN TO NORMALCY 1: IMPLEMENT COMMUNICATIONS PLAN TO KEEP ALL STAKEHOLDERS AND STAFF AWARE OF FACILITY OPERATIONS AND TIMELINES FOR NORMAL SERVICE RESUMTION	COMMAND/PIO

RESOURCES ASSIGNED THIS OPERATIONAL PERIOD			
DIVISION/GROUP NAME	SUPERVISOR	STAFF #s	COMMUNICATIONS
Parking, Sidewalks & Roads Group			
Emergency Power Division			

7. Prepared by   PRINT NAME: _____   SIGNATURE: _____

DATE/TIME: _____   FACILITY: _____

INITIAL INCIDENT ACTION PLAN FORM—LOSS OF INTERNET SERVICE (INCLUDING VOIP PHONES)

1. Incident Name	2. Initial Operational Period - 2 hours

3. Situation Summary — HICS 201 —

There are widespread computer internet connection problems with some computers slow to boot up or not loading. Telephone, pagers and Email is not available and computers used for patient monitoring have been reported as unreliable.

4. Current Incident Management Team (fill in additional positions as appropriate) — HICS 201, 203 —

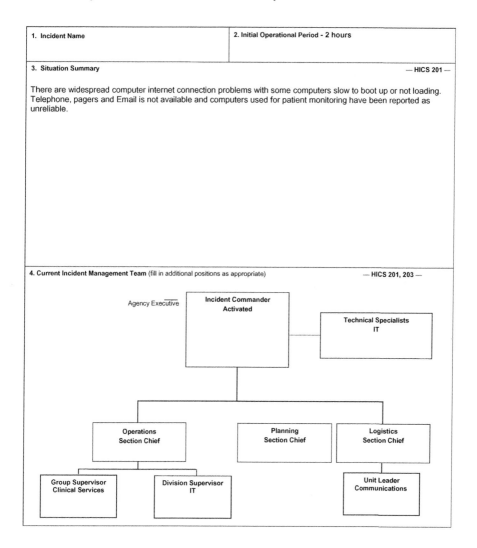

5. Health and Safety Briefing Identify potential incident health and safety hazards and develop necessary measures (remove hazard, provide personal protective equipment, warn people of the hazard) to protect responders from those hazards.

Hazard: Lack of normal emergency alerting systems	Mitigation Strategy: Assign additional safety officers as needed to locations
Hazard:	Mitigation Strategy:
Hazard:	Mitigation Strategy:

6. SMART Incident Objectives – Specific – Measurable – Achievable – Realistic – Timely	Assigned To:
LIFE SAFETY 1: MAINTAIN PATIENT CARE CAPABILITIES – IMPLMENT DOWNTIME PROCEDURES	CLINICAL GROUP
LIFE SAFETY 2: EVALUATE INTERNAL AND EXTERNAL EMERGENCY COMMUNICATION SYSTEMS	COMMO UNIT
LIFE SAFETY 3: PROVIDE FOR SECURITY OF THE HOSPITAL, INCLUDING MANUAL PATROLS AND CONTROLS OF INGRESS AND EGRESS.	OPS
INCIDENT STABILIZATION 1: NOTIFY AFFECTED END USER SUPERVISORY PERSONNEL AND PROVIDE GUIDANCE ON INFORMATION TECHNOLOGY SYSTEMS USE	COMMAND/PIO
INCIDENT STABILIZATION 2: CONDUCT RISK ASSESSMENT OF AFFECTED ENVIRONMENTAL SYSTEMS (E.G., HEATING, VENTILATION, AIR CONDITIONING, AND UTILITIES) AND DEVELOP PLANS TO MAINTAIN AFFECTED SYSTEMS THAT SUPPORT HOSPITAL OPERATIONS.	OPS
INCIDENT STABILIZATION 3: EVALUATE NEED TO LIMIT NON-ESSENTIAL OR ELECTIVE SERVICES	COMMAND/EXEC
PROPERTY/ENVIORNMENTAL CONSERVATION 1:	
RETURN TO NORMALCY 1: ISOLATE AND REPAIR AFFECTED INFORMATION TECHNOLOGY SYSTEMS RESTORE AUTOMATED SYSTEMS AND SERVICES	IT DIVISION

RESOURCES ASSIGNED THIS OPERATIONAL PERIOD			
DIVISION/GROUP NAME	SUPERVISOR	STAFF #s	COMMUNICATIONS
IT Division			
Clinical Group			
Communications Unit			

7. Prepared by PRINT NAME: _____ SIGNATURE: _____

DATE/TIME: _____ FACILITY: _____

INITIAL INCIDENT ACTION PLAN FORM—TORNADO WATCH/ SEVERE WIND EVENT

1. Incident Name	2. Initial Operational Period – 2 hours

3. Situation Summary — HICS 201 —

Your facility has been under a tornado watch and severe weather is predicted to continue for the next several hours with a threat of a tornado warning. Your next shift change is set to occur close to when the worst weather will be in your area. You have activated your IMT to monitor the situation and take pro-active actions.

4. Current Incident Management Team (fill in additional positions as appropriate) — HICS 201, 203 —

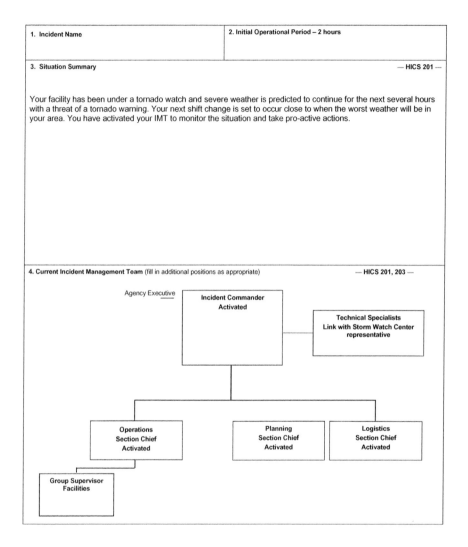

5. **Health and Safety Briefing** Identify potential incident health and safety hazards and develop necessary measures (remove hazard, provide personal protective equipment, warn people of the hazard) to protect responders from those hazards.	
Hazard: Items being picked up by the wind	Mitigation Strategy: Store loose items or secure to solid anchor
Hazard: Breaking glass	Mitigation Strategy: Evaluate exposures, install shields or restrict access
Hazard:	Mitigation Strategy:

6. SMART Incident Objectives – Specific – Measurable – Achievable – Realistic – Timely	Assigned To:
LIFE SAFETY 1: ENSURE STAFF REVIEW UNIT SPECIFIC PROCEDURES IN THE EVENT OF ESCALATION OF THREAT TO TORNADO WARNING	OPS
LIFE SAFETY 2: IMPLEMENT HIGH WIND PROCEDURES, EG. STORE LOOSE ITEMS, TIE DOWNS ATTACHED	FACILITIES GROUP
LIFE SAFETY 3: DISCUSS WITH AGENCY EXECUTIVE OPTIONS OF: A) ASKING NEXT SHIFT TO COME IN EARLY TO PROVIDE EXTRA STAFF AND KEEP THEM OFF THE ROAD DURING WORST WEATHER; B) ASK CURRENT SHIFT TO PREPARE TO STAY OVER IN ONCOMING SHIFT MUST BE DELAYED	AGENCY EXEC
INCIDENT STABILIZATION 1: MONITOR WEATHER CONDITIONS AND SHARE WITH STAFF	COMMAND/PIO
INCIDENT STABILIZATION 2:	
INCIDENT STABILIZATION 3:	
PROPERTY/ENVIORNMENTAL CONSERVATION 1: PREPARE EMERGENCY WINDOW COVERINGS, FLASH FLOOD CONTROL MEASURES	FACILITIES GROUP & LOGISTICS
RETURN TO NORMALCY 1: DISCUSS NEED TO DELAY ELECTIVE PROCEDURES SCEHDULED WITHIN THE NEXT 2-4 HOURS UNTIL THE WEATHER BREAKS.	

RESOURCES ASSIGNED THIS OPERATIONAL PERIOD			
DIVISION/GROUP NAME	SUPERVISOR	STAFF #s	COMMUNICATIONS
Facilities			

7. **Prepared by**   PRINT NAME: _____   SIGNATURE: _____

    DATE/TIME: _____   FACILITY: _____

INITIAL INCIDENT ACTION PLAN FORM—TORNADO WARNING EVENT

1. Incident Name	2. Initial Operational Period – 1 hour

3. Situation Summary — HICS 201 —

Your facility had been under a tornado watch and now it has changed to a tornado warning with a funnel cloud on the ground several miles away and drifting towards your facility. Your IMT has been activated to monitor the situation and take pro-active actions noted in Tornado Watch IAP.

4. Current Incident Management Team (fill in additional positions as appropriate) — HICS 201, 203 —

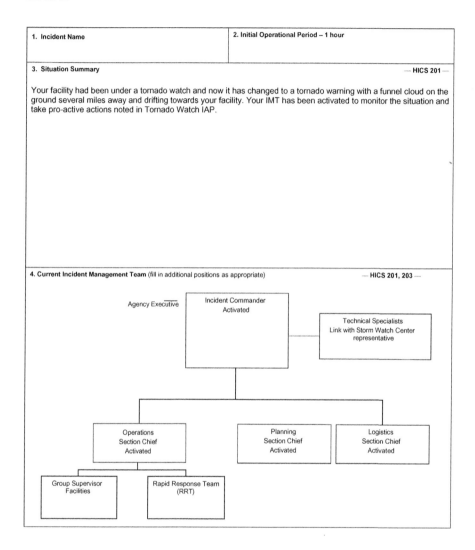

5. Health and Safety Briefing Remind all staff key steps in shelter-in-place protective measures for patients, staff and visitors	
Hazard: Items being picked up by the wind	Mitigation Strategy: Stay indoors and away from glass if possible
Hazard: Breaking glass	Mitigation Strategy: Move patients to indoor hallways and/or away from glass. Utilize blankets, screens or mattresses prn.
Hazard: Unsafe outdoors until front passes	Mitigation Strategy: Lock-down facility to keep everyone inside. ED The only exception. Post staff near exits.

6. SMART Incident Objectives – Specific – Measurable – Achievable – Realistic – Timely	Assigned To:
LIFE SAFETY 1: IMPLEMENT SHELTER-IN-PLACE PROCEDURES	COMMAND/PIO
LIFE SAFETY 2: RESPOND TO FACILITY ISSUES AS REPORTED OR BASED ON HVA PRIORITIES	FACILITIES GROUP
LIFE SAFETY 3: CONDUCT OUTDOOR SWEEP - POST PASSING OF STORM FRONT	OPS/RRT
INCIDENT STABILIZATION 1: DEPLOY RAPID RESPONSE TEAM(S) AS REQUIRED FOR PATIENT CARE ISSUES	OPS/RRT
INCIDENT STABILIZATION 2: CONDUCT DAMAGE ASSESSMENT AFTER STORM PASSES	FACILITIES GROUP
INCIDENT STABILIZATION 3:	
PROPERTY/ENVIORNMENTAL CONSERVATION 1: COVER AT RISK HIGH VALUE EQUIPMENT WITH PLASTIC	FACILITIES GROUP
RETURN TO NORMALCY 1:	

RESOURCES ASSIGNED THIS OPERATIONAL PERIOD			
DIVISION/GROUP NAME	SUPERVISOR	STAFF #s	COMMUNICATIONS
Facilities			
RRT- as many as needed			

7. Prepared by   PRINT NAME: _____   SIGNATURE: _____

DATE/TIME: _____   FACILITY: _____

INITIAL INCIDENT ACTION PLAN FORM—TORNADO EVENT

1. Incident Name	2. Operational Period (#)
	DATE: FROM: _____ TO: _____
	TIME: FROM: _____ TO: _____

3. Situation Summary — HICS 201 —

4. Current Incident Management Team (fill in additional positions as appropriate)   — HICS 201, 203 —

- Incident Commander
  - Technical Specialists
  - Operations Section Chief
    - Group Supervisor
    - Division Supervisor
  - Planning Section Chief
  - Logistics Section Chief

5. Health and Safety Briefing Identify potential incident health and safety hazards and develop necessary measures (remove hazard, provide personal protective equipment, warn people of the hazard) to protect responders from those hazards.

Hazard:	Mitigation Strategy:
Hazard:	Mitigation Strategy:
Hazard:	Mitigation Strategy:

6. SMART Incident Objectives – Specific – Measurable – Achievable – Realistic – Timely	Assigned To:
LIFE SAFETY 1:	
LIFE SAFETY 2:	
LIFE SAFETY 3:	
INCIDENT STABILIZATION 1:	
INCIDENT STABILIZATION 2:	
INCIDENT STABILIZATION 3:	
PROPERTY/ENVIORNMENTAL CONSERVATION 1:	
RETURN TO NORMALCY 1:	

RESOURCES ASSIGNED THIS OPERATIONAL PERIOD

DIVISION/GROUP NAME	SUPERVISOR	STAFF #s	COMMUNICATIONS

7. Prepared by PRINT NAME: _____ SIGNATURE: _____

DATE/TIME: _____ FACILITY: _____

INITIAL INCIDENT ACTION PLAN FORM—LOSS OF WATER EVENT

1. Incident Name	2. Initial Operational Period – 2 hours

3. Situation Summary — HICS 201 —

Without warning your facility has lost municipal water supplies. This was noticed when facets did not work and toilets did not fill back up.

4. Current Incident Management Team (fill in additional positions as appropriate) — HICS 201, 203 —

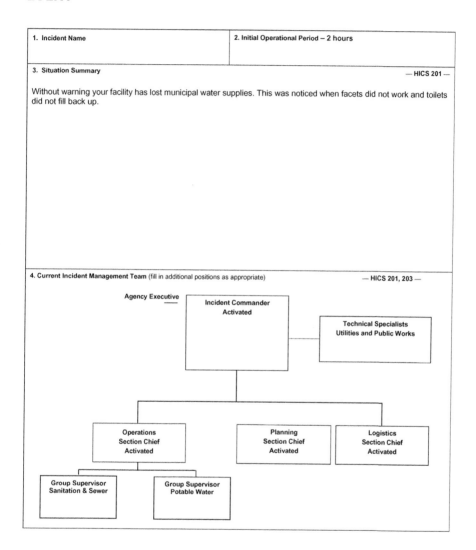

5. Health and Safety Briefing Identify potential incident health and safety hazards and develop necessary measures (remove hazard, provide personal protective equipment, warn people of the hazard) to protect responders from those hazards.

Hazard: Inability to wash hands for sanitation purposes	Mitigation Strategy: Advise staff and visitors to use waterless hand cleaner
Hazard: Consuming non-potable water	Mitigation Strategy: Notify staff not to consume any tap water until confirmed safe when service resumes.
Hazard:	Mitigation Strategy:

6. SMART Incident Objectives – Specific – Measurable – Achievable – Realistic – Timely	Assigned To:
LIFE SAFETY 1: DISTRIBUTE WATERLESS HAND SANITIZER AT ALL WASH BASINS	S & S GROUP
LIFE SAFETY 2: SECURE SOURCE OF WATER FOR REGULAR TOILET FLUSHING	S & S GROUP
LIFE SAFETY 3: SECURE SOURCE OF POTABLE WATER FOR SHORT-TERM CONSUMPTION NEEDS	POTABLE WATER GRP
INCIDENT STABILIZATION 1: CONDUCT ASSESSMENT OF OUTAGE DURATION	PLANNING
INCIDENT STABILIZATION 2: DETERMINE SOURCE OF POTABLE WATER FOR MEDIUM-TERM NEEDS	LOGISTICS
INCIDENT STABILIZATION 3: DETERMINE NEED TO CANCEL NON-EMERGENCY SERVICES	COMMAND/EXEC
PROPERTY/ENVIORNMENTAL CONSERVATION 1:	
RETURN TO NORMALCY 1:	

RESOURCES ASSIGNED THIS OPERATIONAL PERIOD

DIVISION/GROUP NAME	SUPERVISOR	STAFF #s	COMMUNICATIONS
Potable Water Group			
Sanitation and Sewer Group			

7. Prepared by   PRINT NAME: _____   SIGNATURE: _____
                  DATE/TIME: _____   FACILITY: _____

MINI EXERCISE LIST

The following scenarios are designed to take very little time or effort. Copy them on to three by five index cards and hand them out randomly to members of your IMT. I prefer to give them this staff on all three shifts while they are working so they get practice taking off their day to day hat and put on their IMT member hat. It should take them less than 15 min to answer the questions once they start thinking like an IMT member well versed in the Hospital Incident Command System.

Tornado

A tornado warning has been issued with at least one tornado having touched down southwest of our location. It should be passing our location within 20 min and an immediate shelter-in-place order is given by the County Emergency Management Agency (EMA) office.

Will you active the Command Center, yes or no?

Determine:

your first operational period
your initial objectives
draw your first organizational chart and assign names to positions

Flood
Heavier than forecast rains for the past 6 h have created flash food conditions. The County EMA office advises that they expect to see heavy rains continue with significant flooding in low-lying areas and impassable roads. Some employees have already called in, they cannot get in to work.

Will you active the Command Center, yes or no?

Determine:

your first operational period
your initial objectives
draw your first organizational chart and assign names to positions

Hazmat
On the TV in the break room, a special news report comes on about an explosion and subsequent out of control fire at an industrial facility within 5 miles of your facility. Initial reports suggest that an unknown amount of chemicals may have been released and possibly are now burning. Wind direction currently puts your facility upwind.

Will you active the Command Center, yes or no?

Determine:

your first operational period
your initial objectives
draw your first organizational chart and assign names to positions

Loss of Power
The power suddenly is lost at your facility. Staff are used to the generators kicking in and a return with some power within minutes, but 30 min have passed and no power is on. Maintenance reports that a power surge associated with the original outage has fried circuits and they estimate that it will take 6–12 h to repair.

Will you active the Command Center, yes or no?

Determine:

your first operational period
your initial objectives
draw your first organizational chart and assign names to positions

Windstorm

Early in the afternoon and caused by unstable weather, a band of straight-line wind roared through north of the community. Winds of 115 mph were measured. Your power went out but emergency generators kicked in. Several windows were blown out in several out-patient buildings separate from the hospital. Damage reports on the news speak of widespread problems of roads blocked, power out and one major box store collapse with unknown numbers of people inside.

Will you active the Command Center, yes or no?

Determine:

your first operational period
your initial objectives
draw your first organizational chart and assign names to positions

Active Shooter

A man enters the main lobby, steps near a security staff member, and shoots them and the receptionist, both gravely wounded. He moves down a hallway, shooting three more people, killing one before disappearing into the facility. Additional gunshots can be heard. People in the lobby scatter, several calling 9-1-1. A local police officer arrives within minutes, confronts the gunman, and kills him. You have at least six wounded and two dead victims.

Will you active the Command Center, yes or no?

Determine:

your first operational period
your initial objectives
draw your first organizational chart and assign names to positions

Ill Patient

After assessing an elderly febrile adult male patient sitting in the waiting room, it is determined that they were in Saudi Arabia for a week and returned 10 days ago. Chief complaint is cough and shortness of breath.

Will you active the Command Center, yes or no?

Determine:

your first operational period
your initial objectives
draw your first organizational chart and assign names to positions

ADDITIONAL READING

NIMS Implementation Activities for Hospitals and Healthcare Systems, https://www.fema.gov/pdf/emergency/nims/imp_hos.pdf.

Emergency Management Principles and Practices for Health Care Systems—Incident Command System (ICS), Multiagency Coordination Systems (MACS) and the Application of Strategic NIMS Principles, http://www.calhospitalprepare.org/sites/main/files/file-ttachments/empp_unit_2_2nd_edition.pdf.

Proposal to Create the Department of Homeland Security, https://www.dhs.gov/sites/default/files/publications/book_0.pdf.

NIMS Implementation for Healthcare Organizations Guidance, https://www.phe.gov/Preparedness/planning/hpp/reports/Documents/nims-implementation-guide-jan2015.pdf.

The 1993 World Trade Center Bombing: Report and Analysis, https://www.usfa.fema.gov/downloads/pdf/publications/tr-076.pdf.

The 1995 After Action Report Alfred P. Murrah Federal Building Bombing, https://www.ok.gov/OEM/documents/Bombing%20After%20Action%20Report.pdf

THE 9/11 Commission Report—Final Report of the National Commission on Terrorist Attacks Upon the United States—Executive Summary, http://govinfo.library.unt.edu/911/report/911Report_Exec.htm.

Hurricane Katrina—After Action Report—OR-2 DMAT, https://www.hsdl.org/?view&did=766144.

Hurricane Sandy FEMA After—Action Report, https://www.fema.gov/media-library-data/20130726-1923-25045-7442/sandy_fema_aar.pdf.

FEMA—Emergency Management Institute—Online Independent Study Courses, https://training.fema.gov/is/crslist.aspx.

California Emergency Medical Services Authority—Hospital Incident Command System, 2014 Revisions, http://www.emsa.ca.gov/disaster_medical_services_division_hospital_incident_command_system_resources.

Great Tohoku, Japan Earthquake and Tsunami, 11 March 2011, https://www.ngdc.noaa.gov/hazard/11mar2011.html.

FEMA Public Assistance Applicant Handbook, https://www.fema.gov/pdf/government/grant/pa/fema323_app_handbk.pdf.

FEMA Incident Complexity Typing, https://training.fema.gov/emiweb/is/icsresource/assets/incidenttypes.pdf.

USCG Incident Management Handbook, https://www.uscg.mil/d9/D9Response/docs/USCG%20IMH%202014%20COMDTPUB%20P3120.17B.pdf.

Also, available as an App, Search "Coast Guard IMH".

A Better Emergency Initial Incident Action Plan Template, http://abetteremergency.com/blog/wp-content/uploads/2016/08/IAP-Quick-Start.docx.

ASPR Hazard Vulnerability Analysis/Risk Assessment Topic Collection, https://asprtracie.hhs.gov/documents/hazard-vulnerability-risk.pdf.